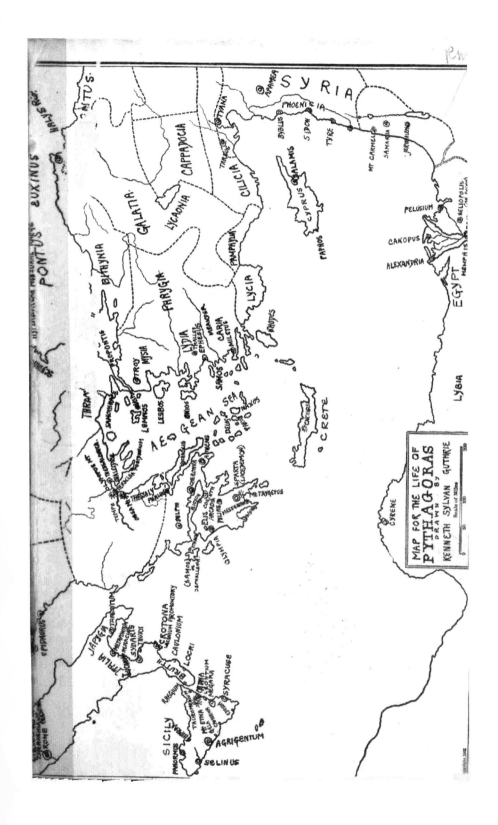

MAP FOR THE LIFE OF
PYTHAGORAS
DRAWN BY
KENNETH SYLVAN GUTHRIE

Scale of Miles

PYTHAGORAS

THE LIFE OF PYTHAGORAS

by Iamblichus of Syrian Chacis

English Version

Edited by

Kenneth Sylvan Guthrie,

Translator of Plotinus.

PLATONIST Press, Box 42,

ALPINE, N.J. U.S.A.

THE PLATONIST PRESS

TEOCALLI, 1177 WARBURTON AVE., YONKERS, N. Y.

553614

INDEX

IAMBLICHUS of Syrian Chalcis's

LIFE   OF   PYTHAGORAS

- - - - - - -

Chapter   I

IMPORTANCE OF THE SUBJECT.

Since wise people are in the habit of invoking the divinities at the beginning of any philosophic consideration, this is all the more necessary on studying that one which is justly named after the divine Pythagoras. Inasmuch as it emanated from the divinities it could not be apprehended without their inspiration and assistance. Besides, its beauty and majesty so surpasses human capacity, that it cannot be comprehended in one glance. Gradually only can some details of it be mastered when, under divine guidance we approach the subject with a quiet mind. Having therefore invoked the divine guidance, and adapted ourselves and our style to the divine circumstances, we shall acquiesce in all the suggestions that come to us. Therefore we shall not begin with any excuses for the long neglect of this sect, not by any explanations about its having been concealed by foreign disciplines, or mystic symbols, nor insist that it has been obscured by false and spurious writings, nor make apologies for any special hindrances to its progress. For us it is sufficient that this is the will of the Gods, which will enable us to undertake tasks even more arduous than these. Having thus acknowledged our primary submission to the divinities, our secondary devotion shall be to the prince and father of this philosophy as a leader. We shall, however, have to begin by a study of his descent and nationality.

CHAPTER II

YOUTH,   EDUCATION,   TRAVELS:

It is reported that Ancaeus, who dwelt in
Cephallenian Samos, was descended from Jupiter,
the fame of which honorable descent might have
been derived from his virtue, or from a certain
magnanimity; in any case, he surpassed the remainder of the Cephallenians in wisdom and renown.
This Ancaeus was, by the Pythian oracle, bidden
form a colony from Arcadia and Thessaly; and besides leading some inhabitants of Athens, Epidaurus
rus, and Chalcis, he was to render habitable an
island, which, from the virtue of the soil and
vegetation was to be called Black-leaved, while
the city was to be called Samos, after Sane, in
Cephallenia. The oracle ran thus: "I bad you,
Ancaeus, to colonise the maritime island of
Sane, and to call it Phyllas." That the colony ari
originated from these places is proved first
from the divinities, and their sacrifices, which
were imported by the inhabitants, second by the
relationships of the families, and third by
their Samian gatherings.
    From the family and alliance of this Ancaeus,
founder of the colony, were therefore decended
Pythagoras's parents Mnesarchus and Pythais. Th··
That Pythagoras was the son of Apollo is a legend
due to a certain Samian poet, who thus described the popular recognition of the nobility
of his birth. Sang he,
        "Pythais, the fairest of the Samian race
         From the embraces of the God Apollo
         Bore Pythagoras, the friend of Jove."
It might be worth while to relate the circumstances of the prevalence of this report. Mnesarchus
had gone to Delphi on a business trip, leaving
his wife without any signs of pregnancy. He enqui
quired of the oracle about the event of his return voyage to Syria, and he was informed that
his trip would be lucrative, and most conformable to his wishes; but that his wife was now
with child, and would present him with a son

who would surpass all who had ever lived in
beauty and wisdom, and that he would be of the
greatest benefit to the human race in everything pertaining to human achievements. But
when Mnesarchus realized that the God, without
waiting for any question about a son, had by an
oracle informed him that he would possess an illustrious prerogative, and a truly divine gift,
he immediately changed his wife's former name
Parthenis to one reminiscent of the Delphic
prophet and her son, naming her Pythais, and
the infant, who was soon after born at Sidon
in Phoenicia, Pythagoras, by this name commemorating that such an offspring had been
promised him by the Pythian Apollo. The asser-
tions of Epimenides, Eudoxus and Xenocrates,
that Apollo having at that time already had
actual connection with Parthenis, causing her
pregnancy, had regularized that fact by predicting the birth of Pythagoras, are by no means
to be admitted. No one will deny that the soul
of Pythagoras was sent to mankind from Apollo's
domain, having either been one of his attendants,
or more intimate associates, which may be inferred both from his birth, and his versatile wisdom.

After Mnesarchus had returned from Syria to
Samos, with great wealth derived from a favorable sea-voyage, he built a temple to Apollo,
with the inscription of Pythius. He took care
that his son should enjoy the best possible education, studying under Creophilus, then under
Pherecydes the Syrian, and then under almost
all who presided over sacred concerns, to whom
he especially recommended his son, that he might
be as expert as possible in divinity. Thus by
education and good fortune he became the most
beautiful and godlike of all those who have been
celebrated in the annals of history.

After his father's death, though he was still
but a youth, his aspect was so venerable, and his
habits so temperate that he was honored and even
reverenced by elderly men, attracting the atten-

tion of all who saw and heard him speak, creating
the most profound impression. That is the reason
that many plausibly asserted that he was a child
of the divinity. Enjoying the privilege of such
a renown, of an education so thorough from
infancy, and of so impressive a natural appearance,
he showed that he deserved all these advantages
by deserving them, by the adornment of piety
and discipline, by exquisite habits, by firmness
of soul, and by a body duly subjected to the
mandates of reason. An inimitable quiet and serenity marked all his words and actions, soaring
above all laughter, emulation, contention, or
any other irregularity or eccentricity; his
influence, at Samos, was that of some beneficent
divinity. His great renown, while yet a youth,
reached not only men as illustrious for their
wisdom as Thales at Miletus, and Bias at Priene,
but also extended to the neighboring cities. He
was celebrated everywhere as the "long-haired
Samian," and by the multitude was given credit
for being under divine inspiration.

When he had attained his eighteenth year,
there arose the tyranny of Policrates; and Pythagoras foresaw that under such a government his
studies might be impeded, as they engrossed the
whole of his attention. So by night he privately
departed with one Hermodamas, — who was surnamed Creophilus, and was the grandson of the
host, friend and general preceptor of the poet
Homer, — going to Pherecydes, to Anaximander
the natural philosopher, and to Thales at Miletus.
He successively associated with each of these
philosophers in a manner such that they all loved
him, admired his natural endowments, and admitted him to the best of their doctrines. Thales
especially, on gladly admitting him to the in-
timacies of his confidence, admired the great
difference between him and other young men, who
who were in every accomplishment surpassed by
Pythagoras. After increasing the reputation
Pythagoras had already acquired, by communicating
to him the utmost he was able to impart to him,
Thales, laying stress on his advanced age and

the infirmities of his body, advised him to go
to Egypt, to get in touch with the priests of
Memphis and Jupiter. Thales confessed that the
instruction of these priests was the source of his
own reputation for wisdom, while neither his own
endowments nor achievements equalled those which
were so evident in Pythagoras. Thales insisted
that, in view of all this, if Pythagoras should
study with those priests, he was certain of becoming the wisest and most divine of men.

CHAPTER III

JOURNEY TO EGYPT

Pythagoras had benefited by the instruction
of Thales in many respects, but his greatest les-
son had been to learn the value of saving time,
which led him to abstain entirely from wine and
animal food, avoiding greediness, confining himself to nutriments of easy preparation and diges-
tion. As a result, his sleep was short, his soul
pure and vigilant, and the general health of his
body was invariable.

Enjoying such advantages, therefore, he sailed to Sidon, which he knew to be his native country, and because it was on his way to Egypt. In
Phoenicia he conversed with the prophets who were
the descendants of Moschus the physiologist, and
with many others, as well as with the local hiero-
phants. He was also initiated into all the mysteries of Byblus and Tyre, and in the sacred functions
performed in many parts of Syria. He was led to
all this not from any hankering after superstition,
as might easily be supposed, but rather from a
desire of and love for contemplation, and from
an anxiety to miss nothing of the mysteries of
of the divinities thith deserved to be learned.

After gaining all he could from the Phoeni-
ian mysteries, he found that they had originated
from the sacred rites of Egypt, forming as it
were an Egyptian colony. This led him to hope that
in Egypt itself he might find monuments of erudition still more genuine, beautiful, and divine.

Therefore following the advice of his teacher
Thales, he left, as soon as possible, through
the agency of some Egyptian sailors, who very
opportunely happened to land on the Phoenician
coast under Mount Carmel, in the temple on the
peak of which Pythagoras for the most part dwelt
in solitude. He was gladly received by the sailors, who intended to make a great profit by sel-
ling him into slavery. But they changed their
mind in his favor during the voyage, when they
perceived the chastened venerability of the mode
of life he had undertaken. They began to reflect
that there was some thing supernatural in the
youth's modesty, and in the manner in which he
had unexpectedly appeared to them on their landing, when, from the summit of Mount Carmel, which
they knew to be more sacred than other mountains,
and quite inaccessible to the vulgar, he had
leisurely descended without looking back, avoiding all delay from precipices or difficult rocks;
and that when he came to the boat, he said nothing
more than, "Are you bound for Egypt?" And farther
that, on their answering affirmatively, he had gone
aboard, and had, during the whole trip, sat silent
where he would be least likely to inconvenience
them at their tasks. For two nights and three
days Pythagoras had remained in the same unmoved
position, without food, drink, or sleep, except
that, unnoticed by the sailors, he might have
dozed while sitting upright. Moreover, the sailors considered that, contrary to their expecta-
tions, their voyage had proceeded without in-
terruptions, as if some deity had been on board.
From all these circumstances they concluded that
a very divinity had passed over with them from
Syria into Egypt. Addressing Pythagoras and each
other with a gentleness and propriety that was u
unwonted, they completed the remainder of their
voyage through a halcyon sea, and at length hap-
pily landed on the Egyptian coast. Reverently
the sailors here assisted him to disembark; and
after they had seen him safe onto a firm beach,
they raised before him a temporary altar, heaped
on it the now abundant fruits of trees, as if

these were the first-fruits of their freight,
presented them to him and departed hastily to
their destination. Pythagoras, however, whose
body had become emaciated through the severity
of so long a fast, did not refuse the sailors'
help in landing, and as soon as they had left
partook of as much of the fruits as was requisite
to restore his physical vigor. Then he went inland,
in entire safety, preserving his wonted tranquility and modesty.

CHAPTER IV.

STUDIES IN EGYPT AND BABYLONIA.

Here in Egypt he frequented all the temples
with the greatest diligence, and most studious
research, during which time he won the esteem and
admiration of all the priests and prophets with
whom he associated. Having most solicitously
familiarized himself with every detail, he did
not, nevertheless, neglect any contemporary ce-
lebrity, whether sage renowned for wisdom, or
or peculiarly performed mystery; he did not fail
to visit any place where he thought he might discover something worth while. That is how he visited all of the Egyptian priests, acquiring all
the wisdom each possessed. He thus passed twen-
ty-two years in the sanctuaries of temples, studying astronomy and geometry, and being initiated
in no casual or superficial manner in all the
mysteries of the Gods. At length, however, he was
taken captive by the soldiers of Cambyses, and
carried off to Babylon. Here he was overjoyed to
associate with the Magi, who instructed him in
their venerable knowledge, and in the most perfect
worship of the Gods. Through their assistance,
likewise, he studied and completed arithmetic,
music, and all the other sciences. After twelve
years, about the fifty-sixth year of his age, he
returned to Samos.

CHAPTER V

TRAVELS IN GREECE; SETTLEMENT AT CROTONA.

On his return to Samos he was recognized by some of the older inhabitants, who found that he had gained in beauty and wisdom, and achieved a divine graciousness; wherefore they admired him all the more. He was officially invited to benefit all men, by imparting his knowledge publicly. To this he was not averse; but the method of teaching he wished to introduce was the symbolical one, in a manner similar to that in which he had been instructed in Egypt. This mode of teaching, however, did not please the Samians, whose attention lacked perseverance. Not one proved genuinely desirous of those mathematical disciplines which he was so anxious to introduce among the Greeks; and soon he was left entirely alone. This however did not embitter him to the point of neglecting or despising Samos. Because it was his home town, he desired to give his fellow-citizens a taste of the sweetness of the mathematical disciplines, in spite of their refusal to learn. To overcome this he devised and executed the following stratagem. In the gymnasium he happened to observe the unusually skilful and masterful ball-playing of of a youth who was greatly devoted to physical culture, but impecunious and in difficult circumstances. Pythagoras wondered whether this youth, if supplied with the necessaries of life, and freed from the anxiety of supplying them, could be induced to study with him. Pythagoras therefore called the youth, as he was leaving the bath, and and made him the proposition to furnish him the means to continue his physical training, on the condition that he would study with him easily and gradually, but continuously, so as to avoid confusion and distraction, certain disciplines which he claimed to have learned from the Barbarians in his youth, but which were now beginning to desert him in consequence of the inroads of the forgetfulness of old age. Moved by hopes of financial support, the youth took up the proposition without delay. Pythagoras then introduced

him to the rudiments of arithmetic and geometry,
illustrating them objectively on an abacus, paying him three oboli as fee for the learning of
every figure. This was continued for a long time,
the youth being incited to the study of geometry
by the desire for honor, with diligence, and in
the best order. But when the sage observed that
the youth had become so captivated by the logic,
ingeniousness and style of those demonstrations
to which he had been led in an orderly way, that
he would no longer neglect their pursuit merely
because of the sufferings of poverty, Pythagoras
pretended poverty, and consequent inability to
continue the payment of the three oboli fee. On
hearing this, the youth replied, that even without the fee he could go on learning and receiving this instruction. Then Pythagoras said, "But
even I myself am lacking the means to procure
food!" As he would have to work to earn his living,
he ought not to be distracted by the abacus and
other trifling occupations. The youth, however,
loth to discontinue his studies, replied, "In
the future, it is I who will provide for you, and
repay your kindness in a way resembling that of
the stork; for in my turn, I will give you three
oboli for every figure." From this time on he
was so captivated by these disciplines, that, of
all the Samians, he alone elected to leave home
to follow Pythagoras, being a namesake of his,
though differing in patronymic, being the son
of Eratocles. It is probably to him that should
be ascribed three books on Athletics, in which
he recommends a diet of flesh, instead of dry
figs, which of course would hardly have been
written by the Mnesarchian Pythagoras.

About this time Pythagoras went to Delos, where
he was much admired as he approached the so-called
bloodless altar of Father Apollo, and wor shipped
it. Then Pythagoras visited all the oracles. He
dwelt for some time in Crete and Sparta, to learn
their laws; and on acquiring proficiency therein
he returned home to complete
        his former omissions.

On his arrival in Samos, he first established
a school, which is even now called, the Semicircle
of Pythagoras, in which the Samians now consult a
about public affairs, feeling the fitness of dis-
pensing justice and promoting profit in the place
constructed by him who promoted the welfare of all
mankind. Outside of the city he formed a cave adapted to the practices of his philosophy, in which
he spent the greater part of day and night, ever
busied with scientific research, and meditating
as did Minos, the son of Jupiter. Indeed he surpassed those who later practised his disciplines
chiefly in this, that they advertised themselves
for the knowledge of theorems of minute importance, while Pythagoras unfolded a complete science of the celestial orbs, founding it on arithmetical and geometrical demonstrations.

Still more than for all this, he is to be admired for what he accomplished later. His philosophy now gained great importance, and his fame
spread to all Greece, so that the best students
visited Samos on his account, to share in his eru-
udition. But his fellow-citizens insisted on employing him in all their embassies, and compelled
him to take part in the administration of public
affairs. Pythagoras began to realize the impossibility of complying with the claims of his country while remaining at home to advance his philosophy; and observing that all earlier philosophers
had passed their life in foreign countries, he determined to resign all political occupations. Besides, according to contemporary testimony, he
was disgusted at the Samians scorn for education.

Therefore he went to Italy, conceiving that
his real fatherland must be the country containing
the greatest number of most scholarly men. Such
was the success of his journey that on his arrival at Crotona, the noblest city in Italy, that
he gathered as many as six hundred followers, who
by his discourses were moved, not only to philoso-
ophical study, but to an amicable sharing of their
worldly goods, whence they derived the names of
Cenobites.

CHAPTER VI

THE PYTHAGOREAN COMMUNITY.

The Cenobites were students that philosophized; but the greater part of his followers were called <u>Hearers</u>, of whom, according to Nicomachus there were qwo thousand that had been captivated by a single oration on his arrival in Italy. These, with their wand children, gathered into one immense auditory, called <u>Auditorium</u>, which was so great as to resemble a city, thus founding a place universally called <u>Greater Greece</u>. This great multitude of people, receiving from Pythagoras laes and mandates as so many divine precepts, without which they declined to engage in any occupation, dwelt together in the greatest general concord, estimated and celebrated by their neighbors as among the number of the blessed, who, as was al ready observed, shared all their possessions.

Such was their reverence for Pythagoras, that they ranked him with the Gods, as a genial beneficent divinity. While some celebrated him as the Pythian, others called him the Northern Apollo. Others considered him Paeon, others, one of the divinities that inhabit the moon; yet others considered that he was one of the Olympian Gods, who, in order to correct and improve terrestrial existence appeared to their contemporaries in human form, to extend to them the salutary light of philosophy and felicity. Never indeed came, nor, for the matter of that, ever will come to mankind a greater good than that which was imparted to the Greeks through this Pythagoras. Hence, even now, the nickname of "long-haired Samian" is still applied to the most venerable among men.

In his treatise on the Pythagoric Philosophy, Aristotle relates that among the principmal arcana of the Pythagoreans was preserved this distinction among rational animals: Gods, men, and beings like Pythagoras. Well indeed may they have done so, inasmuch as he introd-

uced so just and apt a generalization as Gods,
heroes and demons; of the world, of the manifold
motions of the spheres and stars, their oppositions,
tions, eclipses, inequalities, eccentricites
and epicycles; of all the natures contained in
heaven and earth, together with the intermediate
ones, whether apparent or occult. Nor was there,
in all this variety of information, anything
contrary to the phenomena, or to the conceptions
tions of the mind. Besides all this, Pythagoras
unfolded to the Greeks all the disciplines,
theories and researches that would purify he
intellect from the blindness introduced by s
studies of a different kind, so as to enable it
it to perceive the true principles and causes
of the universe.

In addition, the best polity, popular concord
concord, community of possessions among friends,
worship of the Gods, piety to the dead, legislation, erudition, silence, abstinence from
eating the flesh of animals, continence, temperance, sagacity, divinity, and in one word,
whatever is anxiously desired by the scholarly,
was brought to light by Pythagoras.

It was on account of all this, as we have ale
already observed, that Pythagoras was so much
admired.

CHAPTER VII.

ITALIAN POLITICAL ACHIEVEMENTS

Bow we must relate how he travelled, what
places he first visited, and what discourses he
made, on what subjects, and to whom addressed;
for this would illustrate his contemporary
relations. His first task, on arriving in Italy
and Sicily, was to inspire with a love of liberty those cities which he understood had more
or less recently oppressed each other with slavery. Then, by means of his auditors, he liberated
and restored to independence Crotona, Sybaris,
Catanes, Rhegium, Himaera, Agrigentum, Tauromenas, a
and some other cities. Through Charondas the
Catanaean, and Zaleucus the Locrian, he established

laws whichcaused the cities to flourish, and
become models for others in their proximity.
Partisanship, discord and sedition, and that
for several generations, he entirely rooted out,
as history testifies, from all the Italian and
Sicilian lands, which at that time were disturbed by inner and outer contentions. Everywhere,
in private and in public, he would repeat, as
an epitome of his own opinions, and as a per suas-
ive oracle of divinity, that by any means soever,
stratagem, fire, or sword, we should amputate
from the body, disease; from the soul ignorance;
from the belly, luxury; from a city, sedition;
from a household, discord; and from all things
soever, lack of moderation; through which he
brought home to his disciples the quintessence
of all teachings, and that with a most paternal
affection.

For the sake of accuracy, we may state that
the year of his arrival in Italy was that one
of the Olympic victory in the stadium of Eryxidas
of Chalcis, in the sixty-second Olympiad . He became conspicuous and celebrated as soon as he
arrived, just as formerly he achieved instant
recognition at Delos, when he performed his
adorations at the bloodless altar of Father
Apollo.

CHAPTER VIII

INTUITION, REVERENCE, TEMPERANCE, and STUDIOUSNESS

One day, during a trip from Sybaris to Crotona, by the sea-shore, he happened to meet
some fishermen engaged in drawing up from the
deep their heavily-laden fish-nets. He told them
he knew the exact number of the fish they had
caught. The surprised fishermen declared that
if he was right they would do anything he said.
He then ordered them, after counting the fish
accurately, to return them alive to the sea,
and what is more wonderful, while he stood on
the shore, not one of them died, though they

had remained out of their natural element quite
a little while. Pythagoras then paid the fishermen the price of their fish, and departed for
Crotona. The fishermen divulged the occurrence,
and on discovering his name from some children,
spread it abroad publicly. Everybody wanted to se
see the stranger, which was easy enough to do.
They were deeply impressed on beholding his countenance, which indeed betrayed his real nature.

A few days later, on entering in the gymnasium, he was surrounded by a crowd of young men,
and he embraced this opportunity to address them,
exhorting them to attend to their elders, point-
ing out to them the general preeminence of the
early over the late. He instanced that the
east was more important than the west, the morning than the evening, the beginning than the
end, growth than decay; natives than strangers,
city-planners than city-builders; and in general,
that Gods were more worthy of honor than divinities, divinities than semi-divinities, and heroes than men; and that among these the authors
of birth in importance excelled their progeny.
All this, however, he said only to prove by in-
duction, that children should honor their parents,
to whom, he asserted, they were as much indebted f
for gratitude as would be a dead man to him who
should bring him back to life, and light. He
continued to observe that it was no more than
just to avoid paining, and to love preeminently
those who had benefited us first and most.
Prior to the children's birth, these are benefited by their parents exclusively, being the
springs of their offspring's righteous conduct.
In any case, it is impossible for children to err,
by not allowing themselves to be outdistanced
in reciprocation of benefits, towards their
parents. Besides, since from our parents we
learn to honor divinity, no doubt the Gods will
pardon those who honor their parents no less
than those who honor the Gods, (thus making common cause with them). Homer even appled the
paternal name to the King of the Gods, calling
him the father of Gods and men. Many other myth-

.ologists informed us that the chiefs of the Gods
even were anxious to claim for themselves that
superlative affection which, through marriage,
binds children to their parents. That is why
(the Orphic theologians) introduced among the
Gods the terms father and mother, Jupiter begetting Minerva, while Juno produced Vulcan,
the nature of which offspring is contrary,
so as to unite the most remote through friendship.

As this argument about the immortals proved
convincing to the Crotonians, Pythagoras continued to enforce voluntary obedience to the
parental wishes, by the example of Hercules,
who had been the founder of the Crotonian colony.
Tradition indeed informed us that that divinity
had undertaken labors so great nout of obedience to the commands of a senior, and that
after his victories therein, he instituted the
Olympic games in honor of his father. Their
mutual association should never result in hostility to friends, but in transforming their
own hostility into friendship. Their benevolent filial disposition should manifest as
modesty, while their universal philanthropy
should take the form of fraternal consideration
and affection.

Temperance was the next topic of his discourses. Since the desires are most flourishing
during youth, this is the time when control
must be effective. While temperance alone is
universal in its application to all ages, boy,
virgin, woman, or the aged, yet this special
virtue is particularly applicable to youth. Moreover, this virtue alone applied universally to
all goods, those of body and soul, preserving
both the health, and studiousness. This may be
proved conversely. When the Greeks and Barbarians
warred about Troy, each of them fell into the
most dreadful calamities, both during the war,
and the return home, and all this through the
incontinence of a single individual. Moreover
the divinity ordained that the punishment of
this single injustice should last over a thous-
and and ten years, by an oracle predicting the

capture of Troy, and ordering that annually the
Locrians should send virgins into the Temple of
Minerva in Troy.

Cultivation of learning was the next topic
Pythagoras urged upon the young men. He invited
them to observe how absurd it would be to rate
the reasoning power as the chief of their faculties, and indeed consult about all other things
by its means, and yet bestow no time or labor on
its exercise. Attention to the body m ight be
compared to unworthy friends, and is liable to
rapid failure; while erudition lasts till death,
and for some procures post-mortem renown, and
may be likened to good, reliable friends. Pythagoras continued to draw illustrations from history
and philosophy, demonstrating that erudition
enables a naturally excellent disposition to
share in the achievements of the leaders of the
race. For others share in their discoveries by
erudition.

Erudition (possesses four great advantages
over all other goods). First, some advantages,
such as strength, beauty, health and fortitude,
cannot be exercised except by the cooperation of
somebody else. Moreover, wealth, dominion, and
many other goods do not remain with him who imparts them to somebody else. Third, some kinds of
goods cannot be possessed by some men, but all
are susceptible of instruction, according to their
individual choice. Moreover, an instructed man
will naturally, and without any impudence, be
led to to take part in the administration of
the affairs of his home country, (as does not
occur with more wealth). One great advantage of
erudition is that it may be imparted to another
person without in the least diminishing the store
of the giver. For it is education which makes
the difference between a man and a wild beast,
a Greek and a Barbarian, a free man and a slave,
and a philosopher from a boor. In short, erudition
is so great an advantage over those who do not
possess it, that in one whole city and during
one whole Olympiad seven men only were found to

be eminent winners in racing, and that in the
whole habitable globe those that excelled in
wisdom amounted to no more than seven. But in
subsequent times it was generally agreed that
Pythagoras alone surpassed all others in philosophy; for instead of calling himself a sage, he
called himself a philosopher.

CHAPTER IX.

COMMUNITY AND CHASTITY.

What Pythagoras said to the youths in the
Gymnasium, those reported to their elders. Here
upon these latter, a thousand strong, called him
into the senate-hose, praised him for what he
had said to their sons, and desired him to unfold to the public administration any thoughts
advantageous to the Crotonians, which he might
have.

Hid first advice was to build a temple to
the Muses, which would preserve the already
existing concord. He observed to them that all
of these divinities were grouped together
by their common name, that they subsisted only
in conjunction with each other, that they specia
ly rejoiced in social honors, and that(in spit
of all changes) the choir of the Muses subsis
ed always one and the same. They comprehended s
symphony, harmony, rhythm, and all things brood-
ing concord. Not only to beautiful theorems doe

their power extend, but to the general symphon-
ious harmony,
(Justice)was the next desideratum. Their

common country was(not to be victimized selfishly), but to be received as a common deposit
from the multitude of citizens. They should
therefore govern it in a manner such that, as an
hereditary possession they might transmit it in
to their posterity. This could best be effected
if the members of the administration realized

their equality with the citizens, with the only
supereminence of justice. It is from the common
recognition that justice is required in every
place, that were created the fables that Themis is
seatedin the same order with Jupiter, and that
Dice, or rightness, is seated by Pluto, and that
Law is established in all cities, so that whoever i
is unjust in things required of him by his position in society, may concurrently appear unjust
towards the whole world. Moreover, senators should
not make use of any of the Gods for the purpose of
of an oath, inasmuch as their language should be
such as to make them crddible even without any
oaths.

As to their domestic affairs, their government
ment should be the object of deliberate choice.
They should show genuine affection to their own
offspring, remembering that these, from among
all animals, were the only ones who could ap-
preciate this affection. Their associations with
their partners in life, their wives, should be
such as to be mindful that while other compacts
are engraved in tables and pillars, the uxorial
ones are incarnated in children. They should moreov
moreover make an effort to win the affection of thei
their children, not merely in a natural, involuntary manner, but through deliberate choice,
which alone meriterious beneficenve.

He further besought them to avoid connexion
with any but their wives; lest, angered by their
husbands' neglect and vice, these should net
get even by adulterating the race. They should also
also consider that they received their wives from
the Vestal hearth with libations, and brought
them home in the presence of the Gods themselves,
as suppliants would have done. Also that by orderly conduct and temperance they should become models
not only for their family, but also for their
community.

Again, they should minimize public vice, lest
offenders indulge in secret sins to escape the
punishment of the laws, but should rather be impelled to justice from reverence for beauty and
propriety. Procrastination also was to be ended,

inasmuch as opportuneness was the best part
of any deed. The separation of parents from their
children Pythagoras considered the greatest of
evils. While he who is able to discern what is
advantageous to himself may be considered the
best man, next to him in excellence should be
ranked he who can see the utility in what happens to others; while the worst man was he who
waited till he himself was afflicted before before understanding where true advantage lies.
Seekers of honor might well imitate racers, who
do not injure their antagonists, but limit themselves to trying to achieve the victory themselves. Administrators of public affairs should not
betray offense at being contradicted, but on the
other hand benefit the tractable. Seekers of true
gl ory should strive really to become what they
wished to seem; for counsel is not as sacred as
praise, the former being useful only among men,
while the latter mostly referred to the divinities.

In closing, he reminded them that their city
happened to have been founded by Hercules, at a
time when, having been injured by Lacinius, he
drove the oxen through Italy; when, rendering
assistance to Croton by night, mistaking him for
an enemy, he slew him unintentionally. Wherefore
Hercules promised that Wa city should be built
over the sepulchre of Crotonand from him derive
the name Crotona, thus endowing him with immortality. Therefore, said Pythagoras to the rulers
of the city, these should justly render thanks
for the benefits they had received.

The Crotonians, on hearing his words, built a
temple to the Muses, and drove away their
concubines, and requested Pythagoras to address
the young men in the temple of Pythian Apollo,
and the women in the temple of Juno.

CHAPTER X

ADVICE TO YOUTHS

To boys Pythagoras, complying with their parents' request, gave the following advice. They should neither revile any one, nor revenge themselves on those who did. They should devote themselves diligently to learning, which in Greek derives its name from their age. A youth who started out modestly would find it easy to preserve probity for the remainder of his life, which would be a difficult task for one who at that age was not well disposed; nay, for one who begins his course from a bad impulse to run well to to the end is almost impossible.

Pythagoras pointed out that boys were most dear to the divinities; and he pointed out that, in times of great drought, . cities would send boys as ambassadors to implore rain from the Gods, in the persuasion that divinity is especially attentive to children, although such as are permitted to to take part in sacred ceremonies continuously hardly ever arrive at perfect purification. That is also the reason why the most philanthropic of the Gods, Apollo and Love, are, in pictures, universally represented as having the ages of boys. It is similarly recognized that some of the games in which conquerors are crowned were instituted for the behoof of boys; the Pythian, in consequence of the serpent Python having been slain by a boy, and the Nemean and Istimian, because of the death of Archemorus and Melicerta. Moreover, while the city of Crotona was building, Apollo promised to the founder that he would give him a progeny, if he brought a colony into Italy, inferring therefrom that Apollo presided over their development, and that inasmuch as all the divinities protected their age, it was no more than fair that they should render themselves worthy of their friendship.

He added that they should practise hearing,
so that they might learn to speak. Further, that
as soon as they had entered on the path along
which they intebded to proceed for the remainder of their existence, they should imitate
their predecessors, never contradicting those
who were their seniors. For later on, when they
themselves will have grown, they will justly
expect not to be injured by their future juniors.

Because of these moral teachings, Pythagoras
deserved no longer to be called by his patronymic, but that all men should call him divine.

CHAPTER XI

ADVICE TO WOMEN

To the women Pythagoras spake as follows, about sacrifices. To begin with, inasmuch as it was no more than natural that they would wish th that some other person who intended to pray for them should be worthy, nay, excellent, because the Gods attend to those particularly, so also it is advisable that they themselves should most highly esteem equity and modesty, so that the divinities may be the more inclined to grant their requests.

Further, they should offer to the divinities such things as they themselves have with their own hands produced, such as cakes, honey-combs, vapors and perfumes, and should bring them to the altars without the assistance of servants.

They should not worship divinities with blood and dead bodies, nor offer so many things at one time that it might seem they meant never to sacrifi rifice again.

Concerning their association with men, they should remember that their female nature had by their parents been granted the license to love their husbands more excessively than even the authors of their existence. Consequently they should take care neither to oppose their husbands, nor consider that they have subjected their husbands should these latter yield to them in any detail.

It was in the same assembly that Pythagoras is said to have made the celebrated suggestion that, after a woman has had connection with her husband, it is holy for her to perform sacred rites on the same day, which would be inadmissible, had the connection been with any man other than her husband.

He also advised the women that their conversation should always be cheerful, and to endeavor that others may speak good things of them. He further admonished them to care for their good reputation, and to try not to justify the

Fable-writer who accused three women of using
a single eye in common, so great is their mutual willingness to accomodate each other with
the loan of garments and ornaments, without a
witness, when some one of them has special need
thereof, returning them without arguments or
rselitigation.

Further Pythagoras observed that (Mercury)
who is called the wisest of all, who arranged
the human voice, and in short, was the inventor
of names, whether he was a God (in Jupiter,
the supermundane gods, the liberated gods, or
the planet Mercury), or a divinity (the Mercur-
ial order of demons), or a certain divine man
(the Egyptian Thouth, (or in special animals
such as the ibis, ape, or dogs), perceing that
the female sex was most given to devotion, gave
to each of their ages the name of one divinity.
So an unmarried woman was called Core, or Pros-
erpine; a bride, Nympha; a matron, Mother; and
a grandmother, in the Doric dialect, Maia. Consequently, the oracles at Dodona and Delphi are
brought to light by a woman.

By this praise of female piety Pythagoras
is said to have effected so great a change in
popular female attire, that the women no longer
dared to dress up in costly raiment, consecrating thousands of their garments in the temple of
Juno.

fideThis discourse had effect also on marital
fidelity to an extent such that in the Crotonian
region connubial faithfulness became proverbial;
(thus imitating) Ulysses who, rather than abandon Penelope, considered immortality well lost.
Pythagoras encouraged the Crotonian women to
emulate Ulysses, by exhibiting their probity to
their husbands.

In short, through these (social) discourses
Pythagoras acquired great fame both in Crotona,
and in the rest of Italy.

CHAPTER XII

WHY PYTHAGORAS CALLS HIMSELF A PHILOSOPHER
(From Heraclides Ponticus, in 3Cicero.Tusc.v.3)

Pythagoras is said to have been the first to call himself a philosopher, a word which heretofore had not been an appellation, but a description. He likened the entrance of men into the present life to the progression of a crowd to some public spectacle. There assemble men of all descriptions and views. Obe hastens to sell his wares for money and gain; another exhibits his bodily strength for renown; but the most liberal assemble to observe the landscape, the beautiful works of art, the specimens of valor, and the customary literary productions. So also in the present life men of manifold pursuits are assembled. Some are influenced by the desire of riches and luxury; others, by the love of power and dominion, or by insane ambition for glory. But the purest and most genuine character is that of the man who devotes himself to the contemplation of the most beautiful things; and he may properly be called a philosopher.

Pythagoras adds that the survey of the whole heaven, and of the stars that revolve therein, is indeed beautiful, when we consider their order, which is derived from participation in the first and intelligible essence. But that first essence is the nature and number of reasons (or, productive principles), which pervades everything, and according to which all those (celestial) bodies are arranged elegantly, and adorned fittingly. Now veritable wisdom is a science conversant with the first beautiful objects (the intelligibles properly so called); which subsist in invariable sameness, being undecaying and divine, by the participation in which other things also may well be called beautiful. The desire for something like this is philosophy. Similarly beautiful is devotion to erudition; and this notion Pythagoras extended, in order to effect the improvement of the human race.

CHAPTER XIII

HOW ORPHEUS' CONTROL OVER ANIMALS

According to credible historians, his words possessed an admonitory quality that prevailed even with animals, which confirms that, in intelligent men learning tames beasts even wild or irrational. The Daunian bear, who had severely injured the inhabitants, was by Pythagoras detained, long stroking it gently, feeding it on maize and acorns, and after compelling it by an oath to leave alone living beings, he sent it away. It hid itself in the mountains and forest, and was never since known to injure any irrational animal.

At Tarentum he saw an ox feeding in a pasture, where he ate green beans. He advised the herdsman to abstain from this food tell the ox to abstain from this food. The herdsman laughed at him, remarking he did not know the language of oxen; but that if Pythagoras did, he had better tell him so himself. Pythagoras approached the ox's ear, and whispered into it for a long time, whereafter the ox not only refrained from them, but even never tasted them. This ox lived a long while at Tarentum, near the temple of Juno, and was fed on human food by visitors, till very old, considered sacred. Once happening to be talking to his intimates about birds, symbols and prodigies, and observed that all these are messengers of the Gods, sent by them to men truly dear to them, when he brought down an eagle flying over Olympia, which he gently stroked, and dismissed.

Through such and similar occurrences, Pythagoras demonstrated that he possessed the same dominion as Orpheus over savage animals, and that he allured and detained them by the power of his voice.

CHAPTER XIV

PYTHAGORAS'S PREEXISTENCE.

Pythagoras used to make the very best possible approach to men by teaching them what would prepare them to learn the truth in other matters. For by the clearest and surest indications he would remind many of his intimates of the former life lived by their soul before it was bound to their body. He would demonstrate by indibitable arguments that he had once been Euphorbus, son of Panthus, conqueror of Patroclus. He would especially · praise the following funeral Homeric verses pertaining to himself, which he would sing to the lyre most elegantly, frequently repeating them.

> "The shining circlets of his golden hair,
> Which even the Graces might be proud to wear,
> Instarred with gems and gold, bestrew the shore
> With dust dishonored, and deformed with gore.
> As the young olive, in some sylvan scene,
> Crowned by fresh fountains with eternal green,
> Lifts the gay head, in snowy flowerets fair,
> And plays and dances to the gentle air;
> When lo, a whirlwind from high heaven invades
> The tender plant, and withers all its shades;
> It lies uprooted from its genial bed,
> A lovely ruin now defaced and dead;
> Thus young, thus beautiful Euphorbus lay,
> While the fierce Spartan tore his arms away."
> Homer, Iliad, 17, Pope.

We shall however omit the reprts about the shield of this Phrygian Euphorbus, which, among other Trojan spoils was dedicated to the Argive Juno, as being too popular in nature. What Pythagoras, however, wished to indicate by all these particulars was that he knew the former lives he had lived, which enabled him to begin providential attention to others, in which he reminded them of their former existences.

CHAPTER XV

PYTHAGORAS CURED BY MEDICINE AND MUSIC

Pythagoras conceived that the first attention
that should be given to men should be addressed
to the senses, as when one perceives beautiful
figures and forms, or hear beautiful rhythms and
melodies. Consequently he laid down that the first
erudition was that which subsists through music's
melodies and rhythms, and from these he obtained
remedies of human manners and passions, and restored the pristine harmony of the faculties of
the soul. Moreover, he devised me dicines calculated to repress and cure the diseases of both
bodies and souls. There is also, by heavens! something which deserves to be mentioned above all:
namely, that for his disciples he arranged and
adjusted what might be called apparatus and
massage, divinely contriving mingling s of certain diatonic, chromatic and enharmonic melodies,
through which hb easily switched and circulated
the passions of the soul in a contrary direction,
whenever they had accumulated recently, irration-
ally or clandestinely; such as sorrow, rage, pity,
over-emulation, fear, manifold desires, angers,
appetites, pride,   collapse, or spasms. Each
of these he correctrd by the rupe of virtue, at-
tempering them through appropriate melodies, as
if through some salutary medicine.

In the evening, likewise, when his disciples
were retiring to sleep, he would thus liberate
them from the day's perturbations and tumults, pur-
ifyong their intellective powers from the influx-
ive and effluxive waves of corporeal nature, quiet-
ing their sleep, and rendering their dreams pleasing and prophetic. But when they arose again oin
the morning, he would free them from the night's
loginess, coma and torperthrough certain peculiar
chords and modulations, produced by either simply
striking the lyre, or adapting the voice. Not
through instrument or physical voice-organs did
Pythagoras effect this; but by the employment of

a certain indescribable divinity, difficult of apprehension, through which he extended his powers
of hearing, fixing his intellect on the sublime
symphonies of the world, he alone apparently hearing and grasping the universal harmony and consone-
ance of the spheres, and the stars that are moved
through them, producing a melody fuller and more
intense than anything effected by mortal sounds.
This melody was also the result of dissimilar and
varying sounds, speeds, magnitudes and intervals
arranged with reference to each other in a cer
tain musical ratio, producing a convoluted motion
most musical if gentle. Irrigated therefore with
this melody, his intellect ordered and exercised
thereby, he would, to the best of his ability
ex hibit certain symbols of these things to his
 disciples, es pecially through imitations thereof through instruments or the physical organs of
voice. For he conceived that, of all the inhabitants of earth, by him alone were these mundane
sounds understood and heard, as if coming from
the central spring and root of nature. He there
fore thought himself worthy to be taught, and to
learn something about the celestial orbs, and to
be assimilated to them by desire and imitation,
inasmuch as his body alone had been well enough
 thereto conformed by the divinity who had given
birth to him. As to other men, he thought they
should be satisfied with looking to him and the
gifts he possessed, and in being benefited and
corrected through images and examples, in consequence of their inability truly to comprehend the
first and genuine archetypes of things. Just as
to those who are unable to look intently at the
sun we contrive to show its eclipses in either the
reflections of still water, or in melted pitch,
or some smoked glass well burnished brazen mir-ror,
so we spare the weakness of their eyes devis-
ing a method of representing light that is re   'i
flective, though less intense than its archetype,
to those who are interested in this sort of a thin
thing.

This peculiar organization of Pythagoras's
body, far finer than that of any other man,
seems to be what Empodocles was obscurely driving at in his enigmatical verses:

Among the Pythagoreans was a man transcendent
in knowledge;
Who possessed the most ample stores of intellectual wealth,
And in most eminent degree assisted in the
works of the wise.
When he extended all the powers of his intellect,
He easily beheld everything,
As far as ten or twenty ages of the human
race!"

These words "transcendent," "he beheld every
detail of all beings," and "the wealth of intellect," and so on, describe as accurately as at
all possible his peculiar, and exceptionally
accurate method of hearing, seeing and understanding.

CHAPTER XVI

PYTHAGOREAN ASCETICISM

Music therefore performed this Pythagorean soul-adjustment. But another kind of purification of the discursive reason, and also of the whole soul, through various studies, was effected (by asceticism). He had a general notion that disciplines and studies should imply some form of labor; and therefore, like a legislator, he decreed trials of the most varied nature, punishments, and restraints by fire and sword, for innate intemperance, or an ineradicable desire for possession, which the depraved could neither suffer nor sustain. Moreover, his intimates were ordered to abstain from all animal food, and any other that are hostile to the reasoning power by impeding its genuine energies. On them he likewise enjoined suppression of speech, and perfect silence, exercising them for years at a time in the subjugation of the tongue, while strenuously and assiduously investigating and ruminating over the most difficult theorems. Hence also he ordered them to abstain from wine, to be sparing in their their food, to sleep little, and to cultivate an unstudied contempt of, and hostility to fame, wealth, and the like; unfeignedly to reverence those to whom reverence is due, genuinely to exercise democratic assimilation and heartiness towards their fellows in age, and towards their juniorscourtesy, encouragement, without envy.

Moreover Pythagoras is generally acknowleged to have been the inventor and legislator of friendship, under its many various forms, such as universal amity of all towards all, of God towards men through their pietyand scientific theories, or of the mutual interrelation of teachings, or universally of the soul towards the body, and of the rational to the rational part, through philosophy and its underlying theories; or whether it be that of men towards each other, of citizens indeed through sound legislation, but of strangers through a correct physiology; or of the husband to the wife, or of brothers and kindred, through unperverted

communion; or whether, in short, it be of all
things towards all, and still farther, of certain irrational animals through justice, and a
physical connexion and association; or whether
it be the pacification and conciliation of the
body whichof itself is mortal, and of its latent
conflicting powers, through health, and a temperate diet conformable to this, in imitation of
the salubrious condition of the mundane elements,

In short, Pythagoras procured his disciples
the most appropriate converse with the Gods, both
waking and sleeping; something which never occurs
in a soul disturbed by anger, pain, or pleasure,
and surely, all the more, by any base desire,
or defiled by ignorance, which is the most noxious
and unholy of all the rest. By all these inventions, therefore, he divinely purified and healed
the soul, resuscitating and saving its divine
part, and directing to the intelligible its
divine eye, which, as Plato says, is more worth
saving than ten thousand corporeal eyes; for
when it is strengthened and clarified by appropriate aids, when we look through this, we perceive the truth about all beings. In this particular respect, therefore, Pythagoras purified
the discursive power of the soul. This is the
(practical) form that erudition took with him,
and such are the objects of his interest.

CHAPTER XVII

TESTS OF PYTHAGOREAN INITIATION

As he therefore thus prepared his disciples
for culture, he did not immediately receive as
an associate any who came to him for that purpose
until he had tested them and examined them jud-
iciously. To begin with he inquired about their
relation to their parents and ki nsfolk. Next he sur
surveyed their laughter, speech or silence, as to
whether it was unseasonable; further, about their
desires, their associates, their conversation,
how they employed their leisure, and what were
the subjects of their joy or grief. He observed
their form, their gait, and the whole motions of
their body. He considered their frame's natural
indications physiognomically, rating them as
visible exponents of the invisible tendencies
of the soul. Aft

After subjecting a candidate to such trials,
he allowed him to be neglected for three years,
still covertly observing his disposition towards
stability, and genuine studiousness, and whether he
he was sufficiently everse to glory, and ready to
despise popular honors.

After this the candidate was compelled to observe
serve silence for five years, so as to have made
definite experiments in continence of speech, inasmuch as the subjugation of the tongue is the
most difficult of all victories, as has indeed
been unfolded by those who have instituted the
mysteries.

During this probation, however, the property
of each was disposed of in common, being committed to trustees, who were called politicians, econ-
omizers, or legislators. Of these probationers, after
after the quinquennial silence, those who by modest
dignity had won his approval as worthy to share in
his doctrines, then became _esoterics, and within
the veil both heard and saw Pythagoras. Prior
to this they participated in his words through
the hearing alone, without seeing him who remained
within the veil, and themselves offering to him
a specimen of their manners.

If rejected, they were given the double of
the wealth they had brought, but the auditors
raised to him a tomb, as if they were dead;
the disciples being generally called auditors.

Should these later happen to meet the rejected
candidate, they would treat him as a stranger,
declaring that he whom they had by education
modelled had died, inasmuch as the object of
these disciplines had been to be turned out
good and honest men.

Those who were slow in the acquisition of
knowledge were considered to be badly organized,
or, we may say, deficient, and sterile.

If, however, after Pythagoras had studied
them physiognomically, their gait, motions and
state of health, he conceived good hopes of them;
and if, after the five years' silence, and the
emotions and initiations from so many disciplines
together with the ablutions of the soul, and so
many and so great purifications produced by
such various theorems, through which sagacity
and sanctity is ingrained into the soul.........
...........if, after all this even, some one was
found to be still sluggish and dull, they would
raise to such a candidate within the school a
pillar or monument, such as was said to have been
done to Perialus the Thurian, and Cylon the prince
of the Sybarites, who were rejected, they expel-

1   from the auditorium, loading him down with
silver and gold. This wealth had by them been
deposited in common, in the care of certain custodians, aptly called Economics. Should any of
the Pythagoreans later meet with the reject,
they did not recognize him whom they accounted
dead. Hence also Lysis, blaming a certain Hipparchus for having revealed the Pythagorean
doctrines to the profane, and to such as accepted them without disciplines or theory, said,

   " "It is reported that you philosophize indiscriminately and publicly, which is opposed
to the customs of Pythagoras. With assiduity
you did indeed learn them, O Hipparchus; but you
have not preserved them. My dear fellow, you
have tasted Sicilian tit-bits, which you should

not have repeated. If you give them up, I shall
be delighted; but if you do not, you will to me
be dead. For it would be pious to recall the human and divine precepts of Pythagoras, and not to
communicate the treasures of wisdom to those who
have not purified their souls, even in a dream.
It is unlawful to give away things obtained with
labors so great, and with assiduity so diligent
to the first person you meet, quite as much as
to divulge the mysteries of the Eleusynian god-
desses to the profane. Either thing would be unjust and impious. We should consider how long a time
time was needed to efface the stains that had
insinuated themselves in our breasts, before we
became worthy to receive the doctrines of Pythago
goras. Unless the dyers previously purified the
garments in which they wish the desired colors to be
be fixed, the dye would either fade, or be washed
away entirely. Similarly, that divine man prepared
the souls of lovers of philosophy, so that they
might not disappoint him in any of those beautiful
qualities which he hoped they would possess. He
did not impart spurious doctrines, nor stratagems,
in which most of the Sophists, who are at leisure
for no good purpose, entangle young men; but his
knowledge of things human and divine was scientific. These Sophists, however, use his doctrines
as a mere pretext commit dreadful atrocities,
sweeping the youths away as in a drag-net, most
disgracefully, making their auditors become rash
nuisances. They infuse theorems and divine doctrines into hearts whose manners are confused and
agitated, just as if pure, clear water should be
poured into a deep well full of mud, which would
stir up the sediment and destroy the clearness of
the water. Such a mutual misfortune occurs between
such teachers and disciples. The intellect and
heart of those whose initiation has not proceeded
by disciplines, are surrounded by thickets dense
and thorny, which obscure the mild, tranquil and
reasoning power of the soul, and impede the development and elevation of the intellective part.
These thickets are produced by intemperance and
avarice, both of which are prolific.

Intemperance produces lawless marriages,
lusts, intoxications, unnatural enjoyments,
and passionate impulsions which which drive
headlong into pits and abysses. The unbridling
of desires has removed the barriers against
incest with even mothers or daughters, and just
as a tyrant would violate city regulations,
or country's laws, with their hands bound behind
them, like slaves, they have been dragged to the
depths of degradation. On the other hand, avarice
produces rapine, robbery, parricide, sacrilege,
sorcery, and kindred evils. Such being the case,
these surrounding thickets, infested with passions,
will have to be cleared out with systematic disciplines, as if with fire and sword; and when
the reason will have been liberated from so many
and great evils, we are in a position to offer
to it, and implant within it something useful
and good."

So great and necessary was the attention
which, according to Pythagoras, should be paid
to disiplines as introductions to philosophy.

Moreover, inasmuch as he devoted so much
care to the examination of the mental attitudes
oßf prospective disciples, he insisted that the
teaching and communication of his doctrines
should be distinguished by great honor.

CHAPTER XVIII

ORGANIZATION OF THE PYTHAGOREAN SCHOOL

The next step is to to set forth how, after
admission to discipleshipfollowed distribution into
into several classes according to individual merit. As the disciples were naturally dissimilar,
it was impracticable for them to participate in
all things equally, nor would it have been fair for
for some to share in the deepest revelations,
while others might get excluded therefrom, or
others from everything; such discriminations would
being unjust. While he communicated some suitable
portion of his discourses to all, he sought to
benefit everybody; preserving the proportion of
justice, by making every man's merit the index
of the extent of his teachings. He carried this me
method so far as to call som Pythagoreans, and
others Pythagorists, just as we discriminate
poets from poetasters. According to this distinction of names, some of his disciples he considered genuine, and to be the models of the others.
The Pythagoreans' possessions were to be shared
in common, inasmuch as they were to live together, while the Pythagoreists should continue to
manage their own property, though by assembling
frequently they might all be at leisure to pursue the same activities. These two modes of life
which originated from Pythagoras, was transmitted
to his successors.

Among the Pythagoreans there were also two
forms of philosophy, pursued by two classes, the
Hearers and the Students. The latter were universally recognized as Pythagoreans, by all the
rest, though the Students did not admit as much
for the Hearers, insisting that these derived
their instructions not from Pythagoras, but from
Hippasus, who was variously described as either
a Crotonian or Metapontine.

The philosophy of the Hearers consisted in
lectures without demonstrations or conferences
or arguments, merely directing something to be
done in a certain way, unquestioningly preserv-

ing them as so many divine dogmas, non-discus-
sable, and which they promised not to reveal,
esteeming as most wise who more than others
retained them.

Of the lectures there were three kinds; the
first merely announced certain facts; others
expressed what it was especially, and the third,
what should, or should not be done about it.
The objective lectures studied such questions
as, What are the islands of the Blessed? What
are the sun and moon? What is the oracle at Delphi? What is the Tetractys? What is harmony?
What was the real nature of the Sirens? — The
subjective lectures studied the especial nature
of aan object, such as, What is the most just
thing? To sacrifice. What is the wisest thing?
Number. The next wisest is the naming p power.
What is the wisest human thing? Medicine. What
is the most beautiful? Harmony. What is the most
powerful? Mental decision. What is the most excellent? Felicity. Which is the most unquestioned
proposition? Thatt all men are depraved. That is
why Pythagoras was said to have praised the Sal-
aminian poet Hippodomas, for singing:
"Tell, O ye Gods, the source from whence ye came,
 And ye, O Men, how evil ye became."
Such were these subjective lectures, which taught
the distinctive nature of everything.

This sort of study really constitutes the
wisdom of the so-called seven sages. For these
also did not investigate what was good simply,
but especially, nor what is difficult, but what
is particularly so. - - namely, for a man to know
himself. So also they considered not what was
easy, but what was most so, namely, to continue
following out your habits. Such studies resembled, and followed the sages, who however pre-
ceeded Pythagoras.

The practice lectures, which studied what
should or should not be done, considered questions such as, That it is necessary to beget
children, inasmuch as we must leave after us successors who may worship the divinities. Again,
that we should put on first the shoe on the right
foot. That it is not proper to parade on the pub-

lic streets, nor to dip into a sprinkling vessel,
nor to wash in a public bath. For in all these
cases the cleanliness of the agents is uncertain.
Other such problems were, Do not assist maman in
laying down a burden, which encourages him to loit-
er, but to assist him in undertaking something.
Do not hope to beget children from a woman who
is rich. Speak not about Pythagoric affairs without light. Perform libations to the Gods from the
handle of the cup, to make the omen auspicious
and to avoid drinking from the same part (from
which the liquor was poured out?) Wear not the image
image of a God on a ring, for fear of defiling
it, as such resemblances should be protected
in a house. Use no woman ill, for she is a suppliant; wherefore, indeed, we bring her from the
Vestal hearth, and take her by the right hand.
Nor is it proper to sacrifice a white cock, who
also is a suppliant, being sacred to the moon,
and announces the hours. — To him who asks for
counsel, give none but the best, for counsel is
a sacrament. The most laborious path is the
best, just as the pleasurable one is mostly the
worst, inasmuch as we entered into the present
life for the sake of education, which best proceeds
by chastening. — It is proper to sacrifice,
and to take off one's shoes on entering into a
temple. In going to a temple, one should not turn
out of the way; for divinity should not be worshippe
shipped carelessly. — It is well to sustain,
and show wounds, if they are in the breast, but
not if they are behind. — The soul of man incar-
nates in the bodies of all animals, except in
those which it is lawful to kill; hence we should
eat none but those whom it is proper to slay.
Such were subjects of these ethical lectures.

The most extended lectures, however, were
those concerning sacrifices, both at the time
when migrating from the present life, and at
other times; also, about the proper manner of
sepulture.

Of some of these propositions the reason is
assigned; such as, for instance that we must
beget children, to leave successors to worship

the Gods. But no justification is assigned for
the others, although in some cases they are
implied proximately or remotely, such as that
bread is not to be broken, because it contributes to the judgment in Hades. Such merely prob
able reasons, that are additional, are not Pythagoric, but were devised by non-Pythagoreans
who wished to add weight to the statement. Thus
for instance, in respect to the last statement,
that bread is not to be broken, some add the rea
on that we should not (unnecessarily) distribute
what has been assembled, inasmuch as in barbari
times a whole friendly group would together poun
upon a single piece. Others again explain that
precept on the grounds that it is unauspicious,
at the beginning of an undertaking, to make an
omen of fracture or diminution. Moreover, all
these precepts are based on one single underlying
principle, the end of divinity, so that the whol
of every life may result in following God, which
is besides that principle and doctrine of philosophy. For it is absurd to search for good in any
direction other than the Gods. These who do so
resemble a man who, in a country governed by a
king, should do hinor to one of his fellow-citizens who is a magistrate, while neglecting him
Who is the ruler of all of them, Indeed, this is
what the Pythagoreans thought of people who
searched for good elsewhere than from God, For
since He exists as the lord of all things, it mu
be self-evident that good must be requested of
Him alone. For even men impart good to those
they love and enjoy, and do the opposite to thos
they dislike. Such indeed was the wisdom of
those precepts.
There was, however, a certain Aegean named
Hippomedon, one of the Pythagorean Hearers, who
insisted that Pythagoras himself gave the reasons for, and demonstrations of these precepts
himself; but that in consequence of their being
delivered to many, some of whom were slow, the
demonstrations were removed, leaving the bare
propositions. The Pythagorean Students, however,
insist that the reasons and demonstrations were
added by Pythagoras himself, explaining the fif-

f:erence arose as follows. According to them,
Pythagoras hailed from Ionia and Samos, to Italy
then flourishing under the tyranny of Polycrates,
and he attracted as associates the very most prominent men of the city. But the more elderly of
these who were busied with politics, and therefore had no leisure, needed the discourses of
Pythagoras dissociated from reasonings, as they
would have found it difficult to follow his meanings through disciplines and demonstrations,
while nevertheless Pythagoras realized that they
would be benefited by knowing what ought to be
done, even though lacking the underlying reason,
just as physicians' patients obtain their health
without hearing the reasons of every de tail of
the treatment. But Pythagoras conversed through
disciplines and demonstrations with the younger
associates, who were able both to act and learn.
Such then are the differing explanations of the
Hearers and Students.

As to Hippasus, however, they acknowledge that
he was one of the Pythagoreans, but that he met
the doom of the impious in the sea in consequence
of having divulged and explained the method of
squaring the circle, by twelve pentagons; but
nevertheless he obtained the renown of having made
the discovery. In reality, however, this, just
as everything else pertaining to geometry, was
the invention of that man, as they referred to
Pythagoras. But the Pythagoreans say that geometry
was divulged under the following circumstance:
A certain Pythagorean happened to lose his fortune,
to recoup which he was permitted to teach that
science, which, by Pythagoras was called History.

So much then concerning the difference of
each mode of philosophizing, and the classes of
Pythagoras's disciples. For those who heard him
either within or without the veil, and those who
heard him accompanied with seeing, or without seeing him, and who are classified as internal or
external auditors, were none others than these.
Under these can be classified the political,
Economic, and Legislative Pythagoreans.

CHAPTER XIX

ABARIS THE SCYTHIAN

Generally, however, it shoupd be known that
Pythagoras discovered many paths of erudition,
but that he communicated to each only that part
of wisdom which was appropriate to the recipients'
nature and power, of which the following is an
appropriate striking illustration. When Abaris
the Scythian came from the Hyperboreans, he was
already of an advanced age, and unskilled and
uninitiated in the Greek learning. Pythagoras
did not compel him to wade through introductory
theorems, the period of silence, and long auscul-
tation, not to mention other trials, but considered him to be fit for an immediate listener to
his doctrines, and instructed him in the shortest
way, in his treatise on Nature, and one On the God.

This Hyperborean Abaris was elderly, and
most wise in sacred concerns, being a priest of
the Apollo there worshipped. At that time he was
returning from Greece to his country, in order to
consecrate the gold which he had collected to
the God in his temple among the Hyperboreans. As
therefore he was passing through Italy, he saw Pyth
agoras, and identified him as the God of whom he
was the priest.

Believing that Pythagoras resembled to no man,
but was none other than the God himself, Apollo,
both from the venerable associations he saw aroun
him, and from those the priest already knew, he
paid him homage by giving him a sacred dart. This
dart he had taken with him when he had left his
temple, as an implement that would stand him in
good stead in the difficulties that might befall
him in so long a journey: For in passing through
inaccessible places, such as rivers, lakes, marshe
mountains, and the like, it carried him, and by it
he was said to have performed lustrations and ex
pelled winds and pestilences from the cities that
requested him to liberate him from such evils.
For instance, it was said that Lacedemon, after
having been by him purified, was no longer infecte
with pestilence, which formerly had been endemic,

through the miasmatic nature of the ground, in
the suffocating heat produced by the overhanging
mountain Taygetus, just as happens with Cnossus
in Crete. Many other similar circumstances were
reported of Abaris.

Pythagoras, however, accepted the dart, without expressing any amazement at the novelty of
the thing, nor asking why the dart was presented
to him, as if he really was a god. Then he took
Abaris aside, and showed him his golden thigh,
as an indication that he was not wholly mistaken
(in his estimate of his real nature). Then Pythagoras described to him several details of his
distant Hyprborean temple, as proof of deserving
being considered divine. Pythagoras also added
that he came(into the regions of mortality) to
remedy and improve the condition of the human
race, having assumed human form lest men, disturbed by the novelty of his transcendency should
avoid the discipline he advised. He advised Abaris
to stay with him, to aid him in correcting (the
manners and morals) of those they might meet, and
to share the common resources of himself and his
associates, whose reason led them to practice
the precept that <u>the possessions of friends are
common</u>.

So Abaris stayed with him, and was compendi-
ously taught physiology and theology; and instead
of divining by the entrails of beasts, he revealed to him the art of prognosticating by numbers,
conceiving this to be a method purer, more divine,
and more kindred to the celestial numbers of the
Gods. Also he taught Abaris other studies for
which he was fit.

Returning however to the purpose of the present treatise, Pythagoras endeavored to correct
and amend different persons according to their
individual abilities. Unfortunately most of these
particulars have neither been publicly transmitted,
nor is it easy to describe that which has been
transmitted to us concerning him.

CHAPTER XIX.
PSYCHOLOGICAL REQUIREMENTS

We must now set forth a few of the most celebrated points of the Pythagoric discipline,
and land-marks of their distinctive studies.

When Pythagoras tested a novice, he considered the latter's ability to hold his counsel,
"echemuthein" being his technical term for this.
namely, whether they could reserve and preserve
what they had heard and learned. Next, he examine.
their modesty, for he was much more anxious that
they should be silent, than that they should
speak. Further, he tested every other quality,
for instance, whether they were astonished by
the energies of any immoderate desire or passion.
His examination of their affectibility by desire
or anger, their contentiousness or ambition,
their inclination to friendship or discord, was
by no means superficial. If then after an accurate survey these novices were approved as of
worthy manners, he then directed his attention
to their facility in learning, and their memory.
He examined their ability to follow what was
said, with rapidity and perspicuity; and then,
whether they were impelled to the disciplines
taught them by temperance and love. For he laid
stress on natural gentleness. This he called
<u>culture</u>. Ferocity he considered hostile to such
a kind of education. For savage manners are
attended by impudence, shamelessness, intemperance, sloth, stupidity, licentiousness, disgrace,
and the like, while their opposites attend mildness and gentleness.

These things then he considere. in making trial
of those that came to him, and in these the Learner
Learners were exercised. Those that were adapted
to receive the goods of the wisdom he possessed
he admitted to discipleship; endeavoring to elev-
ate them to scientific knowledge; but if he perceived that any novice was unadapted to them,
he expelled him as a stranger and a barbarian.

(In the original the XXth chapter continues until after the second next paragraph).

CHAPTER XX

DAILY PROGRAM

   The studies which he delivered to his associate
tes, ware as follows; for those who committed themselves to the guidance of his doctrine acted thus.
   They took solitary morning walks to places whoch
happened to be appropriately quit, to temples or
groves, or other suitable places. They thought it
it inadvisable to converse with any one until
they had gained inner serenity, focussing their
reasoning powers; they considered it turbulent
to mingle in a crowd as soon as they rose from
bed; and that is the reason why these Pythagoreans
always selected the most sacred spots to walk.
   After their morning walk they associated
with each other, especially in temples, or, if
this was not possible, in similar places. This time
time was employed in the discussion of disciplines
and doctrines, and in the correction of manners.
   (Chapter XX) After an association so holy
they turned their attention to the health of
the body. Most of them were rubbed down, and
raced; fewer wrestled, in gardens or groves; others
others in leaping with leaden weights on their
hands, or in oratorical gesticulations, with
a view to the strengthening of the body, studiously selecting for this purpose opposite exercises.
   They lunched on bread and honey, or on the
honey-comb, avoiding wine. Afterwards, they
held receptions to guests and strangers, conformably to the mandates of the laws, which
was restricted to this time of day.
   In the afternoon they once more betook
themselves to walking, yet not alone, as in the
morning walk, but in parties of two or three,
rehearsing the disciplines they had learned, and
exercising themselves in attractive studies.
   After the walk, they patronized the bath; and afte
after whose ablution they gathered in the common dining-room, which accommodated no more
than a group of ten. Then were performed liba-

tions and sacrifices, with fumigations and incense. Then followed supper, which closed before
the setting of the sun. They ate herbs, raw and
boiled, maize, wine, and every food eatable with
bread. Of any animals lawful to immolate, they
ate the flesh; but they rarely partook of fish,
which was not useful to them, for certain causes
Animals not naturally noxious were neither to
be injured, nor slain. This supper was followed
by libations, succeeded by readings. The young-
est read what the eldest advised, and as they
  suggested.

When they were about to depart, the cup-
bearer poured out a libation for them, after
which the eldest would announce precepts, such a
as the following: That a mild and fruitful plant
should neither be injured nor corrupted, nor
any harmless animal. Further, that we should
speak piously, and form suitable conceptions of
divine, tutelary and heroic beings, and similarly, of parents and benefacters. Also, that we
should aid, and not obstruct the enforcement of
laws. Whereafter, all separated, to go home.

They wore a white garment, that was pure.
They also lay on white and pure beds, the coverlets of which were made of linen, not wool. The
They did not hunt, not undertake any similar
exercise. Such were the precepts daily delivered
to the disciples of Pythagoras, in respect to
eating and living.

43

CHAPTER XXII

Tradition tells of another kind of teaching, by Pythagorean maxims pertaining to human opinions and practices, some examples of which may here be mentioned. It advised to remove strife firdm from true friendship. If possible, this was to apply to all friendship; but at all events to that towards parents, elders, and benefactors. Existing friendships with such as these would bnot be preserved (but destroyed) by rivalry, contention, anger, and subsequent graver passions. The scars and ulcers which their advice sometimes cause should be minimized as much as possible, which will be effected if especially the younger of the two should learn how to yield, and subdue his angry emotions. On the other hand, the so-called "paedartases," or corrections and admonitions of the elder towards the younger, should be made with much suavity of manners, and great caution; also with much solicitude and tact, which makes the reproof all the more graceful and useful.

Faith should never be separated from friendship, whether seriously or in jest. Existing friendship cannot survive the insinuation of deceit between professors of friendship.

Nor should friendship be affected by misfortune or other human vicissitude; and the only rejection of friendship which is commendable is that which follows definite and incurable vice.

Such is an example of the Pythagorean hortatory maxims, which extended to all the virtues, and the whole of life.

CHAPTER XXIII

USE OF PARABLES IN INSTRUCTION

Pythagoras considered most necessary the
use of parables in instruction. Most of the Greeks
had adppted it, as the most ancient; and it had been
been both preferentially and in principle employed by the Egyptians, who had developed it in the
most varied manner. In harmony with this it will
be found that Pythagoras attended to it sedulous-
ly, if from the Pythagoric symbols we unfold
their significance and arcane intentions, devel-
oping their content of rectitude and truth, libe.-
erating them from their enigmatic form. When,
according to straightforward and uniform tradition they are accommodated to the sublime intelligence of these philosophers, they deify
beyond human conception.

Those who came from this school, not only
the most ancient Pythagoreans, but also those who
during his old age were still young, such as Phi-
lolaos, and Eurytus, Charondas and Zaleucus,
Brysson and the elder Archytas, Aristaeus, Lysis
and Empdocles, Zamlxis and Epimenides, Milo and
Leucippus, Alcmaeon and Hippasus, and Thymaridas
were all of that age, a multitude of savants, in-
comparably excellent, — all these adopted this
mode of teaching, both in their conversations,
and commentaries and annotations. Their writings
also, and all the books which they published,
most of which have been preserved to our times,
were not composed in popular or vulgar diction,
or in a manner usual to all other writers, so as
to be immediately understood, but in a way such

as to be not easily apprehended by their readers.
For they adopted Pythagoras's law of reserve, in
an arcane manner concealing divine mysteries
from the uninitiated, obscuring their writings
and mutual conversations.

The result is that they who presents theses
symbols without unfolding their meaning by a
suitable exposition, runs the danger of exposing

them to the charge of being ridiculous and inane,
trifling and garrulous. When however they are
expounded according to these symbols, and made
clear and obvious even to the crowds, then they
will be found analogous to prophetic sayings,
such as the oracles of the Pythian Apollo. Their
admirable meaning will inspire those who unite
intellect and scholarliness.

It might be well to mention a few of them,
  explain this mode of discipline.
Not negligently enter into a temple,
 Nor adore carelessly, even if only at the doors.
Sacrifice and adore unshod.
 Shunning public roads, walk in unfrequented paths.
 Not without light speak about Pythagoric affairs.

Such is a sketch of the symbolic mode of
teaching adopted by Pythagoras.

CHAPTER XXIV

DIETARY SUGGESTIONS

Since food, used properly and regularly, greatly contributes to the best discipline, it may be interesting to consider Pythagoras's precepts on the subject. Forbidden was generally all food causing flatulence or indigestion, while he recommended the contrary kind of food that preserve and are astringent. Therefore he recommended the nutritious qualities of millet. Rejected was all food foreign to the Gods, as withdrawing us from communion with them. On the other hand, he forbade to his disciples all food that was sacred, as too honorable to subserve common utility. He exhorted his disciples to abstain from such things as were an impediment to prophecy, or to the purity and chastity to the soul, or to the habit of temperance, and virtue. Lastly, he rejected all things that were an impediment to sanctity, and disturbed or obscured the other purities of the soul, and the phantasms which occur in sleep. Such were the general regulations about food.

Specially, however, the most contemplative of the philosophers, who had arrived at the summit of philosophic attainments, were forbidden superfluous   , food such as wine, or unjustifiable food, such as was animated; and not to sacrifice animals to the Gods, nor by any means to injure animals, but to observe most solicitous justice towards them. He himself lived after this manner, abstaining from animal food, and adoring altars undefiled with blood. He was likewise careful to prevent others from destroying animals of a nature kindred to ours, and rather corrected and instructed savage animals, than injured them as punishment. Further, he ordered abstaining from animal food even to politicians; for as they desired to act justly to the highest degree, they must certainly not injure any kindred animals. How indeed could

they persuade others to act justly, if they themselves were detected in an insatiable avidity in
devouring animals allied to us? These are conjoined to us by a fraternal alliance through the
communion of life, and the same elements, and the
commingling of these. Eating of the flesh of certain animals was however permitted to those
whose life was not entirely purified, philosophic
and sacred; but even for these was appointed a
definite time of abstinence. Besides, these were
not to eat the heart, nor the brain, which were
entirely forbidden to all Pythagoreans. For these
organs are predominant, and as it were ladders
and seats of wisdom and life.

Food other than animal was by him also considered sacred, on account of the nature of divine reason. Thus his disciples were to abstain
from mallows, because this plant is the first messenger and signal of the sympathy of celestial
with terrestrial natures. Moreover, the fish
melanurus was interdicted because sacred to the
terrestrial gods. Likewise, the _erythinus_. Beans
physical, psychic and sacred on account of many causes
also were interdicted, on account of many causes,
physical, psychic and sacred.

Many other similar precepts were enjoined,
in the attempt to lead men to virtue through
their food.

CHAPTER XXV.

MUSIC AND POETRY.

Pythagoras was likewise of opinion that music, if properly used, greatly contributed to health. For he was wont to use it in no careless way, but as a purification. Indeed, he restricted this word to signify music used as medicine.

About the vernal season he used a melody in this manner. In the middle was placed a person who played on the lyre, and seated around him in a circle were those able to sing. Then the lyrist in the centre struck up, and the singers raised certain paeans, through which they were evidently so overjoyed that their manners became elegant and orderly. This music instead of medicines was also used at certain other times.

Certain melodies were devised as remedies against the passions of the soul, as also against despondency and gnashing of the teeth, which were invented by Pythagoras as specifics. Further he employed other melodies against anger and rage, and all other aberrations of the soul. Another kind of modulation was invented against desires. He likewise used dancing, which was accompanied by the lyre, instead of the pipe, which he conceived to have an influence towards insolence, being theatrical, and by no means liberal. For the purpose of correcting the soul, he also used select verses of Homer and Hesiod.

It is related among the deeds of Pythagoras that once, through a spondaic song, he extinguished the rage of a Tauromenian lad, who, after feasting by night, intended to burn the vestibule of the house of his mistress, on seeing her issuing from the house of a rival. (To this rash attempt the lad had been inflamed by a Phrygian song, which however Pythagoras at once suppressed. As Pythagoras was astronomizing he happened to meet this Phrygian piper at an unseasonable time of night, and persuaded him to change his Phrygian song for a spondaic one; through which the fury of the lad being immediately repressed, he return-

ed home in an orderly manner, although but a lit
le while since he had stupidly insulted Pythagoras on meeting him, would bear no admonition, and
could not be restrained.

Here is another instance. Anchitus, the host
of Empedocles, had, as judge, condemned to death
the father of a youth, who rushed on Anchitus with
drawn sword, intending to slay him. Empedocles
changed the youth's intention by singing, to
his lyre, that vers of Homer (Od.4):

"Nepenthe, without gall, o'er every ill
Oblivion spreads; —"
thus saving his host Anchitus from death, and
the youth from committing murder. It is said
that from that time on the youth became one of
the most faithful disciples of Pythagoras.

The Pythagoreans distinguished three states
of mind, called exartysis, or readiness; synarmo-
ge, or fitness, and epaphe, or contact, which converted the soul to contrary passions, and these
could be produced by certain appropriate songs.

When they retired, they purified their reasoning powers from the noises and perturbations to
which they had been exposed during the day, by
certain odes and hymns which produced tranquil
sleep, and few, but good dreams. But when they
arose from slumbers, they again liberated themselves from the dazedness and torpor of sleep
by songs of another kind. Sometimes the passions
of the soul and certain diseases were, as they
said, genuinely lured by enchantments, by musical sounds alone, without words. This is indeed
probably the origin of the general use of this
word epode.

Thus therefore through music Pythagoras produced the most beneficial correction of manners
and lives.

CHAPTER XXVI

THEORETICAL MUSIC (from Nicomachus).

While describing Pythagoras's wisdom in instructing his disciples, we must not fail to note
that he invented the harmonic science and ratios.
But to explain this we must go a little backwards
in time. Once as he was intently considering
music, and reasoning with himself whether it
would be possible to devise some instrumental
assistance to the sense of hearing, so as to
systematize it, as sight is made precise by
· · · · · · · · · · · · · · · · · · ·
the compass, rule, and telescope, or touch is
made reckonable by balance and measures, —
so thinking of these things Pythagoras happened to pass by a brazier's shop, where providentially he heard the hammers beating out a
piece of iron on the anvil, producing sounds
that harmonized, except one. But he recognized
in these sounds the concord of the octave, the
fifth, and the fourth. He saw that the sound
between the fourth and the fifth, taken by itself, was a dissonance, and yet completed the
greater sound among them. Delighted, therefore,
to find that the thing he was anxious to discover had by divine assistance succeeded, he
went into the smithy, and by various experiments discovered that the difference of sound
arose from the magnitude of the hammers, but
not from the force of the strokes, nor from
the shape of the hammers, nor from the change
of position of the beaten iron. Having then
accurately examined the weights and the swing
of the hammers, he returned home, and fixed
one stake diagonally to the walls, lest some
difference should arise from there being several of them, or from some difference in the
material of the stakes. From this stake he
then suspended four gut-strings, of similar
materials, size, thickness and twist. A weight
was suspended from the bottom of each. When
the strings were equal in length, he struck

two of them simultaneously, he reproduced the former
former intervals, forming different pairs. He
discovered that the string stretched by the greats
est weight, when compared with that stretched by the
the smallest weight, the interval of an octave.
The weight of the first was twelve pounds, and
that of the latter, six, Being therefore in a doubl
double ratio, it formed the octave, which was
made plain by the weights themselves. Then he found
found that the string from which the greatest
weight was suspended compared with that from
which was suspended the weight next to the smallest, and which weight was eight pounds, profuced the interval known as the fifth. Hence he discovdiscovered that this interval is in a ratio of
one and a half to one, or three to two, in which
ratio the weights also were to each other. Then
he found that the string stretched by the greatest weight produced, when conpared with that
which was next to it, in weight, namely, nine
pounds, the interval called the fourth, analogous
to the weights. This ratio, therefore, he discovered to be in the ratio of one and a third to
one, or four to three; while that of the string
from which a weight of nine pounds was suspended,
to the string which had the smallest weight, again
in a ratio of
three to two, which is 9 to 6. In like manner,
the string next to that from which the smallest
weight was suspended, was to that which had the
smallest weight, in the ratio of 4 to 3 (being 8 to 6
to 6). but to the string which had the greatest
weight, in a ratio of 3 to 2, being 12 to 8.
Hence that which is between the fifth and the
fourth, and by which the fifth exceeds the fourth,
is proved to be as fiine is to eight. But either
way it may be proved that the octave is a system
consisting of the fifth in conjunction with the
fourth, just as the double ratio consists of three t
to two, and four to three; as for instance 12,
8 and 6; or, conversely, of the fourth and the
fifth, as in the double ratio of four to three
and three to two, as for instance, 12, 9 and 6.
Thus therefore, and in this order, having con

formed both his hand and hearing to the suspended weights, and having established according to
them the ratio of the proportions, by an easy
artifice he transferred the common suspension
of the strings from the diagonal stake to the
head of the instrument which he called "chordo-
tonon," or string-stretcher. Then by the aid of
pegs he priduced a tension of the strings analogous to that effected by the weights.

Employing this method, therefore, as a basis,
and as it were an infallible rule, he afterwar s
extended the experiment to other instruments,
namely, the striking of pans, to pipes and r: ,
to monochords, triangles, and the like, in all c.
which he found the same ratio of numbers to o -
tain. Then he named the sound which participates
in the number 6, tonic; that which participates
of the number 8, and is four to three, subdom-
inant; that which participated of the number 9,
and is one tone higher than the sub-dominant,
he called flominant, and 9 to 8; but that which
participates of the number 12, octave.

Then he filled up the middle spaces with
analogous sounds in diatonic order, and formed
an octochord from symmetric numbers: from the
double, the three to two, the four to three,
and from the difference of these, the 8 to 9.
Thus he discovered the harmonic progression,
which tends by a certain physical necessity
from the lowest to the most acute sound, diaton-
ically.

Later, from the diatonic he progressed to the
chromatic and enharmonic orders, as we shall lat :r
show when we treat of music. This diatonic scale,
however, seems to have the following progression:
a semi-tone, a tone, and a tone; and this is the
fourth, being a system consisting of two tones,
and of what is called a semi-tone. Afterwards,
adding another tone, we produce the fifth, which
is a system consisting of three tones and a semi-
tone. Next to this is the system of a semi-tone,
a tone, and a tone, forming another fourth,
that is, another four to three ratio. Thus in the
more ancient octave indedd, all the sounds from
the lowest pitch which are with respect to each

other fourths, produce everywhere with each other
fourths; the semitone, by transition, receiving
the first, middle and third place, according to
that tetrachord. Now in the Pythagoric octave,
however, which by conjunction is a system of the
tetrachord and pentachord, but if disjoined is
a system of two tetrachords separated from each
other, the progression is from the gravest to
the most acute sound. Hence all sounds that by
their distance from each other are fifths, with
each other produce the interval of the fifth;
the semi-tone successively proceeding into four
places, the first, second, third, and fourth.
This is the way in which music was said to have
been discovered by Pythagoras. Having reduced it
to a system, he delivered it to his disciples
to utilize it to produce things as beautiful
as possible.

(The story of the smithy is an ancient error,
as pieces of iron give the same note whether
struck by heavy or light hammers. Pythagoras
may therefore have brought the discovery with him
from Egypt, though he may also have developed the
further details mentioned in this chapter).

CHAPTER XXVII.

MUTUAL POLITICAL ASSISTANCE.

Many deeds of the Pythagoreans in the political sphere are deservedly praised. At one time the Crotonians were in the habit of making funerals and internments too sumptuous. Thereupon one of them said to the people that once he had heard Pythagoras converse about divine natures, during which he had observed that the Olympian divinities attended to the dispositions of the sacrificers, and not to the multitude of the offerings. The terrestrial gods, on the contrary, as being interested in less important matters, rejoiced in lamentations and banquets, libations, delicacies, and obsequial pomp; and as proof thereof, the divinity of Hades is called Pluto, from his wish to receive. Those that honor him slenderly (he does not much care for), and permits to stay quite a little while; but he hastens to draw down those disposed to spend profusely on funeral solemnities, that he may obtain the honors offered in commemoration of the dead. The result was that the Crotonians that heard this advice were persuaded that if they conducted t themselves moderately in misfortunes, they would be promoting their own salvation, but would die prematurely if immoderate in such expenses.

A certain difference arose about an affair in which there was no witness. Pythagoras on A Pythagorean was made arbitrator; and he led both litigants to a certain monument, announcing that the man buried was exceedingly equitable. The one prayed that he might receive much reward for his good life, while the other declared that the defunct was no better off for his opponent's prayers. The Pythagorean condemned the latter, confirming that he who praised the dead man for his worth had earned credibility.

In a cause of great moment, this Pythagorean decided that one of the two who had agreed to settle that affair by arbitration, should pay four talents, while the other should receive two. Then from him who had received two he took three, and gave them to the other, so that each had been mulcted one talent (the text is confused).

Two persons had fraudulently deposited a garment with a woman who belonged to a court of justice, and told her that she was not to give it to either of them unless both were present. Later, with intent to defraud, one claimed and got the common deposit, saying he had the consent of the other party. The other one turned informer and related the compact made at the beginning to the magistrates. A certain Pythagorean, however, as arbitrator, decided that the woman was guiltless, construing the claimed assent as constructive presence.

Two persons, who had seemed to be great friends, but who had gotten to suspect each other through calumnies of a sycophant, who told the one the other had taken undue liberties with his wife. A Pythagorean happened to enter the smithy where the injured party was finding fault with the blacksmith for not having sufficiently sharpened a sword he had brought him for that purpose. The Pythagorean suspecting the use to which the sword was to be put said, "The sword is sharper than all things except calumny." This caused the prospective avenger to consider that he should not rashly sin against his friend who was within on an invitation (for the purpose of killing him).

A stranger in the Temple of Aesculapius accidentally dropped his belt, on which were gold ornaments. When he tried to pick it up, he was informed that the temple-regulations forbad picking up anything on the floor. He was indignant, and a Pythagorean advised him to remove the golden ornaments which were not touching the floor, leaving the belt which was. (Text corrupt).

During a public spectacle, some cranes flew over the theatre. One sailor said to his companion, "Do you see the witnesses?" A Pythagorean near by haled them into a court presided over by a thousand m agistrates, where, being examined they confessed to having thrown certain boys into the sea, who, on drowning had called on the cranes, flying above them, to witness to the deed. This story is mistakenly located elsewhere, but it really happened at Crotona.

Certain recent disciples of Pythagoras were at variance with each other, and the junior came to the senior, declaring there was no reason to refer the matter to an arbitrator, inasmuch as all they needed to do was to dismiss their anger. The elder agreed, but regretted he had not been the first to make that proposition.

We might relate here the story of Damon and Phinthias, of Plato and Archytas, and of Clinias and Prorus. At present however we shall limit ourselves to that of Eubulus the Messenians, who, when sailing homeward, was taken captive by the Tyrrhenians, where he was recognized by a Pythagorean named Nausithus, who redeemed him from the pirates, and sent him home in safety.

When the Carthaginians were about to send five thousand soldiers into a desert island, the Carthaginian Miltiades saw among them the Argive Pythagorean Possiden. Approaching him, and without revealing his intentions, he advised him to return home with all possible haste. He placed him in a ship then sailing near the shore, supplied him with the travel necessaries, and thus saved him from the impending danger.

He who would try to relate all the fine deeds that beautified the mutual relations of the Pythagoreans would find the task exceeding space and patience. I shall therefore pass on to show that some of the Pythagoreans were competent administrators, adapted to rule. Many were custodians of the laws, and ruled over certain Italian cities, infolding to them, and advising them to adopt the most salutary measures, while themselves refusing all pay. Though greatly calumniated, their probity and the desire of the

citizens prevailed to make them administrators.
At this time the best governed states seem to
have been in Italy and Sicily. One of the best
legislators, Chatondas the Catanean, was a Pythagorean, and so were the celebrated Locrian legislators Zaleucus and Timares. Pythagoreans also
were those Rheginic polities, called the Gymnas-ar
iarchic, named after Theocles. Excelling in studies and manners which were then adopted by their
fellow-citizens, were Phytius, Theocles, Elecaon,
and Aristocrates. Indeed, it is said that Pythagoras was the originator of all political erudition, when he said that nothing existent is pure,
inasmuch as earth participates of fire, fire of
air, and air of water, and water of spirit. Likewise the beautiful participates in the deformed,
the just of the unjust, and so on; so that from
this principle human impulse may (by proper direction) be turned in either direction. He also sai
said that there were two motions, one of the body,
which is irrational, and one of the soul, which i
is the result of deliberate choice. He also said
polities might be likened to three lines whose
extremities join, forming a (triangle containing)
one right angle (the lines being as 4, 3 and 2;
so that one of them is as 4 to 3, another as 3
to 2, and the other (3) is the arithmetical
medium between 2 and 4. Now when, by reasoning,
we study the mutual relations of these lines,
and the places under them, we shall find that they
represent the best image of a polity. Plato
plagiarized, for in his Republic he clearly says,
"That the result of the 4 to 3 ratio, conjoined
with the 5 ratio, produces two harmonies." (This
means that) he cultivated the moderation of the
passions, and the middle path between extremes,
rendering happy the life of his disciples by
relating them to ideals of the good.

We are also told that he persuaded the Crotonians to give up associations with courtesans and prostitutes. Crotenian wives came to Deine
the wife of the Pythagorean Brontinus, who was
a wise and splendid woman, the author of the maxim that it was proper for women to sacrifice on

the same day they had risen from the embraces
of their husbands), —(which some ascribe to
Pythagoras's wife Theano), — and entreated ..r
to persuade Pythagoras to discourse to them on
their continence as due to their husbands, This
she did, and Pythagoras accordingly made an address to the Crotonians, which successfully ended the then prevalent incontinence.

When ambassadors came from Sybaris to Crotona
to demand the (return of) the exiles, and
Pythagoras, seeing one of the ambassadors who
with his own hand had slain one of Pythagoras's
friends, made no answer whatever. But when this
man insisted on an explanation, and addressed
Pythagoras, the latter said it was unlawful to
converse with murderers. This induced many to
believe he was Apollo.

All these stories, together with what we
mentioned above about the destruction of tyrants,
and the democratization of the cities of Italy an
and Sicily, and many other circumstances, are
eloquent of the benefits conferred on mankind
by Pythagoras, in political respects.

CHAPTER XXVIII.

DIVINITY OF PYTHAGORAS.

Henceforward we shall confine ourselves to the works flowing from Pythagoras's virtues. As usual, we shall begin from the divinities, endeavoting to exhibit his piety, and marvellous deeds. Of his piety, let this be a specimen: that he knew what his soul was, whence it came into the body, and also its former lives, of this giving the most evident indications. Again, once passing over the river Nessus along with many associates, he addressed the river, which, in a distinct and clear voice, in the hearing of all his associates, answered, "Hail, Pythagoras!"

Further, all his biographers insist that during the same dayhe was present in Metapontum in Italy, and at Tauromenium in Sicily, discoursing with his disciples in both places, although these cities are separated, both by land and sea by many stadia, the traveling over which consumes many days.

It is also a matter of common report that he showed his golden thigh to the Hyperborean Abaris, who said that he resembled the Apollo worshipped among the Hyperboreans, and of whom Abaris was the priest; and that he had done this so that Abaris might be certified thereof, and that he was not deceived therein.

A myriad of other more admirable and divine particulars are likewise unanimouslu and uniformly related of the man, such as infallible predictions of earthquakes, rapid expulsions of pestilences, and hurricanes, instantaneous cessations of hail, tranquilizations of the waves of rivers and seas, in order that his disciples might the more easily pass over them. The power of effecting miracles of this kind was achieved by Empedocles of Agrigentum, Epimenides the Cretan, and Abaris the Hyperborean, and these perrm: formed them in many places. Their deeds were so manifest that Empedocles was surnamed a windstiller, Epimenides an expiator, and Abaris an air-walker, because, carried on the dart given him by the Hyperborean Apollo, he passed over rivers, and seas and inaccessible places like one carried on air. Many think that Pythagoras did the same thing, when in the same day he discoursed with his disciples at Metapontum and Tauromenium. It is also said that he predicted there would be an earthquake from the water of a well which he had tasted; and that a ship was sailing with a prosperous wind, would be submerged in the sea. These are sufficient proofs of his piety.

Pitching my thoughts on a higher key, I wish to exhibit the principle of the worship of the Gods, established by Pythagoras and his disciples: that the mark aimed at by all plans, whether to do or not to do, is consent with the divinity. The principle of their piety, and indeed their whole life is arranged with a view to follow God. Their philosophy explicitly asserts that men act ridiculously in searching for good from any source other than God; and that in respect the conduct of most men resembles that of a man who, in a country governed by a king should reverence one of the city magistrates, neglecting him who is the ruler of all of them. Since God exists as the lord of all things, it is evident and acknowledged that good must be requested of him. All men impart good to those they love, and admire, and the contrary to those they dislike. Evidently we should do those things in which God delights. Not easy, however, is it for a man to know which these are, unless he obtains this knowledge from one who has heard God, or has heard God himself, or procures it through divine art. Hence also the Pythagoreans were studious of divination, which is an interpretation of the benevolence of the Gods. That such an employment is worth while will be admitted by one who believes in the Gods; but he who thinks that either of these is folly will also be of opinion that both are foolish. Many of the precepts of the Pythagoreans were derived from the mysteries, the Hyperboreans

WHICH were not the fruits of arrogance, in their
estimation, but derived from divinity.

Indeed, Pythagoreans give full belief to myth-
ological stories such as are related of Aristeas
the Proconesian, and Abaris the Hyperborean, and
such like. To them every such thing seems credible, and worthy of being tried out. They also
frequently recollect apparently fabulous particulars, not disbelieving anything which may be referred to the divinity. For instance, it is said
that the Pythagorean Eurytus, disciple of Philolaus, the Pythagorean related that a shepherd
feeding his flock near Philolaus's tomb heard someone singing. His interlocutor, instead of disbelieving the story, asked what kind of harmony it
was. Again, a certain person told Pythagoras that
he once seemed to be conversing with his deceased fa
father, in his dreams, and asked Pythagoras what
this might signify. The answer was, Nothing; even
though the conversation with his father was genuine."As therefore, said he, nothing is signified
by my conversing with you, neither is anything
signified by your conversing with your father."

In all these matters they considered that
the stupidity lay with the sceptics, rather than
with themselves; for they do not conceive that
some things, and not others, are possible with
the Gods, as fancy the Sophists; they thought
that with the Gods all things are possible. This
very assertion is the beginning of some verses
attributed to Linus:

"All things may be the objects of our hopes,
 Since nothing hopeless anywhere is found;
 All things with ease Divinity effects,
 And naught can frustrate his almighty power."

They thought that their opinions deserved
to be believed, because he who first promulgated
them was not some chance person, but a divinity.
This indeed was one of their pet puzzlers: "What
was Pythagoras?"   For they say that

the Hyperborean Apollo, of which this was an indication that rising up, while at the Olympian
games, he showed his golden thigh; and also that
he received the Hyperborean Abaris as his guest,
and by him was presented with the dart on which
he rode through the air. But it is said that
this Abaris came from the Hyperborean regios
to collect gold for his temple, and that he predicted a pestilence. He also dwelt in temples, and
was never seen to eat or drink. It is likewise
said that rites (of his) are performed by the
Lacedemonians, and that on this account Lacedemon is never infested with pestilence. Pythagoras
therefore caused this Abaris to acknowledge (that
he was more than man), refeiving from him at the
same time the golden dart, without which it was
not possible for him to find his way. In Metapontum also also, certain persons praying that they
might obtain what a ship contained that was sailing into port, Pythagoras said to them, You will
then have a dead body. In Sybaris, too, he caught
a deadly serpent, and a shaggy one too, and drove
it away. In Tyrrhenia also he caught a small serpent, whose bite was fatal. In Crotona it is said
that a white eagle allowed Pythagoras to stroke
it. When a certain person wished to hear him
converse, Pythagoras said it was impossible until
some sign appeared. Later a white bear was seen
in Cauconia, whose death he deaclared to a person
who came to announce to him its death. He likewise reminded Myllias the Crotonian that he had
formerly lived as Midas the son of Gordius, and
Myllias journeyed to Asia to perform at the sepulchre of Midas such rites as Pythagoras had commanded him. The person who purchased Pythagoras's
residence dug up what had been buried in it, did
not dare to tell any one what he saw (on this occasion.) Although he did not suffer (any divine
vengeance) for this offence, he was seized and
executed for the sacrilege of taking a golden
beard that had fallen from a statue. These stories
and other such are by the Pythagoreans related
lend authority to their opinions. As their verid-
icity is generally acknowledged, and as they could
not possibly have happened to a mere man, they consequently think it is clear that the stories about

Pythagoras should be received as referring not to
to a mere man, but to a super-man. This is also
what is met 'by their maxim, that <u>man, bird, and
another thi   thing are bipeds,</u> thereby referring
to Pythagoras. Such, therefore, on account of his
piety, was Pythagoras; and such he was truly though
thought to be.

Oaths were religiously observed by the Pythagoreans, who were mindful of that precept of
theirs,

"As duly by law, thy homage pay first to the
                    immortal Gods;
Then to thy oath, and last to the heroes
                    illustrious."

For instance: A Pythagorean was in court, and
asked to take an oath. Rather than to disobey
this principle, although the oath would have been
a religiously permitted one, he preferred to pay
to the defendant a fine of three talents.

Pythagoras taught that no occurrence happened by chance or luck, but rather conformably to
divine Providence, and especially so to good and
pious men. This is well illustrated by a story
from Androcides's treatise on Pythagoric symbols,
about the Tarentine Pythagorean Thymaridas.He
was happening to be sailing away from his country,
his friends were all present to bid him farewell,
and to embrace him. He had already embarked when
some one cried to him, "O Thyramidas, I pray that
the Gods may shape all your circumstances according to your wishes!" But he retorted, "Predict me
better things; namely, thatwhat may happen to me
may be conformable to the will of the Gods!" For
he thought it more scientific and prudent to
not to resist or grumble against divine providence.

If asked about the source whence these men
derived so much piety, we must acknowledge that
the Pythagorean number-theology was cleary fore-
shadowed, to some extent, in the Orphic writings.
Ner is it to be doubted that when Pythagoras composed his treatise Concerning the Gods, he received assistance from Orpheus, wherefore indeed
that theological treatise is sub-titled, the

learned and trustworthy Pythagoreans assert, by
Telauges; taken from the commentaries left by
Pythagoras himself to his daughter Damo, Telauges'
sister, and which, after her death, were said
to have been given to Bitale, Damo's daughter, and
and to Telauges, the son of Pythagoras, and husband of Bitale, when he was of mature age, for he
was, at Pythagoras's death left very young with
his mother Theano. Now wh can judge who it was that
that delivered what there is said of the Gods, from
from the Sacred Discourse, or Treatise on the
Gods, which bears both titles. For we read:
    "Pythagoras, the son of Mnesarchus was in-
structed in what pertains to the Gods when he
celebrated orgies in the Thracian Libethra,
being therein initiated by Aglaophemus; and
that Orpheus, the son of Calliope, having learned
wisdom from his mother in the mountain Pangaeus,
said that the eternal essence of number is the
most providential principle of the universe,
of heaven and earth, and of the intermediate
nature; and farther still, that it is the root
of the permanency of divine natures, of Gods,
and divinities."
    From this it is evident that he learned from
the Orphic writers that the essence of the Gods
is defined by number. "Through the same numbers
also, he produced a wonderful prognostication
and worship of the Gods, both o f which are particularly allied to numbers."
    As conviction is best produced by an objective
fact, the above principle may be proved as
follows. When Abaris performed sacred ritesaccord-
ing to his customs, he procured a fore-knowledge
of events, which is studiously cultivated by all
the Barbarians, by sacrificing animals, especially birds; for they think that the entrails of such
such animals are particularly adapted to this
purpose. Pythagoras, however, not wishing to
suppress his ardent pursuit of the truth, but to
guide it into a safer way, without blood and
slaughter, and also because he thought that a
cock was sacred to the sun,"furnished him with
a consummate knowledge of all truth, through
arithmetical science." Frompiety, also, he

rived faith concerning the Gods. For Pythagoras always insisted that nothing marvellous
concerning Gods or divine teachings should be
disbelieved, inasmuch as the Gods are competent
to effect anything. But the divine teachings
in which we must believe are those delivered by
Pythagoras. The Pythagoreans therefore assumed
and believed what they taught (on the a priori
ground that) they were not the offspring of
false opinion. Hence Eurytus the Crotonian, the
disciple of Philolaus, said that a shepherd
feeding his sheep near Philolaus's tomb had
heard some one singing. But the person to whom
this was related did not at all question this,
merely asking what kind of harmony it was. Pythagoras himself also, being asked by a certain
person the significance of converse with his
defunct father in sleep, answered that it
meant nothing. For neither is anything portend-
ed by your speaking with me, said he.

Pythagoras wore clean white garments, and
used clean white coverlids, avoiding the woolen
ones. This custom he enjoined on his disciples.

In speaking of super-human natures, he used
honorable appellations, and words of good omen,
on every occasion mentioning and reverencing the
Gods; so, while at supper, he performed libations
to the divinities, and taught his disciples
daily to celebrate the super-human beings with
hymns. He attended likewise to rumors and omens,
prophecies and lots, and in short to all unexpected circumstances. Moreover, he sacrificed
to the Gods with millet, cakes, honey-combs,
and fumigations. But he did not sacrifice animals, nor did any of the contemplative philosophers. His other disciples, however, the
Hearers and the Politicians, were by him ordered
to sacrifice animals such as a cock, or a lamb,
or some other young animal, but not frequently;
but they were prohibited from sacrificing oxen.

Another indication of the honor he paid the
Gods was his teaching that his disciples must
never use the names of the divinities uselessly
in swearing. For instance, Syllus, one of the

Crotonian Pythagoreans, paid a fine rather than
swear, though he could have done so without vi-
olating the truth. Just as the Pythagoreans abstained from using the names of the Gods, so
also, through reverence, they were unwilling to
name Pythagoras, indicating him whom they meant
by the invention of the Tetraktys. Such is the
form of an oath ascribed to them:
"I swear by the discoverer of the Tetraktys,
            which is the spring of all our wisdom;
The perennial fount and root of Nature."

In short, Pythagoras imitated the Orphic
mode of writing, and (pious) disposition, the
way they honored the Gods, representing them in
images and in brass not resembling our (human for
form), but the divine receptacle (of the sphere),
because they comprehend and provide for all things
being of a nature and form similar to the universe.

But his divine philosophy and worship was
compound, having learned much from the Orphic
followers, but much also from the Egyptian priests
the Chaldeans and Magi, the mysteries of Eleusis,
Imbrus, Samothracia, and Delos and even the Celtic
and Iberian. It is also said that Pythagoras's
Sacred Discourse is current among the Latins,
not being read to or by all, but only by those
who are disposed to learn the best things, avoiding all that is base.

He ordered that libations should be made
thrice, observing that Apollo delivered oracles
from the tripod, the triad being the first number.
Sacrifices to Venus were to be made on the sixth
day, because this number is the first to partake
of every number, and when divided in every possible way, receives the power of the numbers
subtracted, and those that remain. Sacrifices to
Hercules, however, should be made on the eighth
day, of tje month, counting from the beginning,
commemorating his birth in the seventh month.

He ordained that those who entered into a
temple should be clothed in a clean garment, in
which no one had slept; because sleep, just as
black and brown, indicates sluggishness, while

cleanliness is a sign of equality and justice
in reasoning.

If blood should be found unintentionally spil-
led in a temple, there should be made a lustra-
tion, either in a golden vessel, or with sea-
water; gold being the most beautiful of all
things, and the measure of exchange of everything else; while the latter was derived from
the principle of moistness, the food of the first
and more common matter. Also, children should not bo
be brought forth in a temple; where the divine
part of the soul should not be bound to the
body. On a festal day neither should the hair be
cut, not the nails pared; as it was unworthy to
disturb the worship of the Gods to attend to our
own advantage. Nor should lice be killed in a
temple, as divine power should not participate
in anything superfluous or degrading.

The Gods should be honored with cedar, laurel,
cypress, oak and myrtle; nor should the body be
purified with these, nor any of them be cut with
the teeth.

He also ordered that what is boiled should n
not be roasted, signifying hereby that mildness
has no need of anger.

The bodies of the dead he did not suffer to
be burned, herein following the Magi, being unwilling that anything (so) divine (as fire)
should be mingled with mortal nature. He thought
it holy for the dead to be carried out in white
garments; thereby obscurely prefiguring the simple
and first nature, according to number, and the
principle of all things.

Above all, he ordained that an oath should
be taken religiously; since that which is behind
(the futurity of punishment) is long.

He said that it was much more holy to be
{ an   -   injured than to kill a man;
for judgment is pronounced in Hades, where the
soul and its essence, and the first nature of
things is correctly appraised.

He ordered that coffins should not be made
of cypress, either because the sceptre of Jupiter was made of this wood, or for some other

mystic reason.

Libations were to be performed before the altar of Jupiter the Savior, of Hercules, and the Dioscuri; thus celebrating Jupiter as the presiding cause and leader of the meal; Hercules as the power of Nature, and the Dioscuri, as the symphony of all things. Libations should not be offered with closed eyes, as nothing beautiful should be undertaken with bashfulness and shame.

When it thundered, one ought to touch the earth, in remembrance of the generation of things.

Temples should be entered from places on the right hand, and left from the left hand; for the right hand is the principle of what is called the odd number, and is divine; while the left hand is a symbol of the even number, and of dissolution.

Such are many of the injunctions he is said to have adopted in the pursuance of piety. Other particulars which have been omitted may be infer- red from what has been given. Hence the subject may be closed.

CHAPTER XXIX.

SCIENCES AND MAXIMS.

The Pythagoreans' Commentaries best express
his wisdom; being accurate, concise, savoring of
the ancient elegance of style, and deducing the
conclusions exquisitely. They contain the most
condensed conceptions, and are diversified in
form and matter. They are both accurate and eloquent, nfull of clear and indubitable arguments, accompanied by scientific demonstration,
in syllogystic form; as indeed will be discovered
by any careful reader.

In his writings, Pythagoras , from a supernal
source, delivers the science of intelligible natu-
ures and the Gods. Afterwards, he teaches the
whole of physics, completely unfolding ethics and
logic. Then come various disciplines and other
excellent sciences. There is nothing pertaining
to human knowledge which is not discussed in
these encyclopedic writings. If therefore it is
acknowledged that of the (Pythagoric) writings
which are now in circulation, some were written
by Pythagoras himself, while others consist of
what he was heard to say, and on this account
are anonymous, though of Pythagoric origin; —
if all this be so, it is evident that he was
abundantly skilled in all wisdom.

It is said that while he was in Egypt he
very much applied himself to geometry. For
Egyptian life bristles with geometrical problems;
since, from remote periods, when the Gods were
fabupously said to have reigned in Egypt, on
account of the rising and falling of the Nile,
the skilful have been compelled to measure all
the Egyptian land which they cultivated; where-
from indeed the science's name, geometry, was
derived. Besides, the Egyptians studied the
theories of the celestial orbs, in which Pythagoras also was skilled. All theorems about lines
seem to have been derived from that country.

All that relates to numbers and computation
is said to have been discovered in Phoenicia.
The theorems about the heavenly bodies have by
some been referred to the Egyptians and Chaldeans in common. Whatever Pythagoras received,
however, he developed further, he arranged them
for learners, and personally demonstrated them
with perspicuity and elegance. He was the first
to give a name to philosophy, describing it as
as a desire for and love of wisdom, which latter
he defined as the science of objectified truth.
Beings he defined as immaterial and eternal natures, alone possessing a power that is efficacious, as are incorporeal essences. The rest of
things are beings only figuratively, and considered such only through the participation of real
beings; such are corporeal and material forms,
which arise and decay without ever truly existing. Now wisdom is the science of things which
are truly beings; but not of the mere figurative
entities. Corporeal natures are neither the objects of science, nor admit of a stable knowledge, since they are infinite, and by science in-
comprehensible, and when compared with universals
resemble non-beings, and are in a genuine sense
non-definable. Indeed it is impossible to conceive that there should be a science of things
not naturally the objects of science; nor could
a science of non-existent things prove attractive to any one. Far more desirable will be things
which are genuine beings, existing in invariable
permanency, and always answering to their description. For the perception of objects existing only
figuratively, never truly being what they seem
to be, follows the apprehension of real beings,
just as the knowledge of particulars is posterior to the science of universals. For, as said
Archytas, he who properly knows universals will
also have a clear perception of the nature of
particulars. That is why beings are not simple,
only-begotten, nor simple, but various and multiform. For those genuine beings are intelligible
and incorporeal natures, while others are corporeal, falling under the perception of sense, com-

municate with that which is really existent only
by participation. Concerning all these Pythagoras
as formed sciences the most apposite, leaving not
thing uninvestigated. Besides, he developoed the
master-sciences of method, common to all of them,
such as logic, definitions, and analysis, as may
be gathered from the Pythagoric commentaries.

To his intimates he was wont to utter symbolically oracular sentences, wherein the smallest number of words were pregnant with the most
multifarious significance, not unlike certain
oracles of the Pythian Apollo, or like nature
herself in tiny seeds, the former exhibiting conceptions, and the latter effects innumerable in
multitude, and difficult to understand. Such was
Pythagoras's own maxim, "The beginning is the half
of the whole." In this and similar utterances
the most divine Pythagoras concealed the sparks
of truth, as in a treasury, for those capable of
being kindled thereby. In this brevity of diction
he deposited an extension of theory most ample,
and difficult to grasp, as in the maxim, "All
things accord in number," which he frequently
repeated to his disciples. Another one was,
"Friendship ie equality; equality is friendship."
He even used single words, such as "cosmos," or,
adorned world; or, "philosophy!" or further,
"tetractys!"

All these and many other similar inventions
were by Pythagoras devised for thebenefit and
amendment of his associates; and by those that
understood them they were considered to be so
worthy of veneration , and so divinely inspired,
that those who dwelt in the common auditorium
adopted this oath:
"I swear by the discoverer of the Tetraktys,
          which is the spring of all our wisdom;
The perennial fount and root of Nature."
This was the form of his so admirable wisdom.

Of the sciences honored by the Pythagoreans
not the least were music, medicine and divination.
....Of medicine, the most emphasized part was
dietetics; and they were most scrupulous in its
exercise. First they sought to understand the

physical symptoms of equanimity, labor, fasting
and repose. They were nearly the first to make
a business of the preparation of food, and to
describe its methods. More frequently than their
predecessors the Pythagoreans used poultices,
however disapproving more of medicated ointments,
which they chiefly limited to the cure of ulcer-
ations. Most they disapproved of cuts and cauter-
izations. Some diseases they cured by incantations.
Music, if used in a proper manner, was by Pythagoras supposed to contribute greatly to health.
The Pythagoreans likewise employed select sentences of Homer and Hesiod for the amendment of
souls.

The Pythagoreans were habitually silent and
prompt to hear, and he won praise who listened
(most effectively). But that which they had lea
learned and heard was supposed to be retained
and preserved in the memory. Indeed this ability
of learning and remembering determined the amount
of disciplines and lectures, inasmuch as learning
is the power by which knowledge is obtained,
and remembering that by which it is preserved.
Hence memory was greatly honored, abundantly exercised, and given much attention. In learning
also it was understood that they were not to
dismiss what they were taught, till its first
rudiments had been entirely mastered; This was
their method of recalling what they daily heard.
No Pythagorean rose from his bed till he had
first recollected the transactions of the day before; and he accomplished this by endeavoring to
remember what he first said, or heard, or ordered
done by his domestics before rising; or what was
the second or third thing he had said, heard or
commanded. The same method was employed for the
remainder of the day. He would try to remember
the identity of the first person he had met on
leaving home, and who was the second; and with
whom he had discoursed first, second or third.
So also he did with everything else, endeavoring
to resume in his memory all the events of the
whole day, and in the very same order in which
each of them had occurred. If however after ris-

ias there was enough leisure to do so, the Pythagorean reminisced about day before yesterday.
Thus they made it a point to exercise their
memories systematically; considering that the ability
ability of remembering was most important for
experience, science and wisdom.

This Pythagorean school filled Italy with
philosophers; and this place which before was
unknown, was later, on account of Pythagoras
called Greater Greece, which became famous for
its philosophers, poets and legislators. Indeed
the rhetorical arts, demonstrative reasonings and
legislation was entirely transferred from Greece.
As to physics, we might mention the principal
physiologists, Empedocles and the Elean Parmenides. As to ethical maxims, there is Epicharmus,,
whose conceptions are used by almost all philosephers.
ophers.

Thus much concerning the wisdom of Pythagoras, how in a certain respect he very much impelled all his hearers to its pursuit, so far as
they were adapted to its participation, and how
perfectly he delivered it.

CHAPTER XXX.

JUSTICE AND POLITICS

How he cultivated and delivered justice to humanity we shall best understand if we trace it to its first principle, and ultimate cause. Also we must investigate the ultimate cause of injustice, which will show us how he avoided it, and what methods he adopted to make justice fructify in his soul.

The principle of justice is mutuality and equality, through which, in a way most nearly approximating union of body and soul, all men become cooperative, and distinguish the mine from the thine, as is also testified by Plato, who learned this from Pythagoras. Pythagoras effected this in the best possible manner, by erasing from common life every thing private, while increasing everything held in common, so far as ultimate possessions, which after all are the causes of tumult and sedition. (Among his disciples everything was common, and the same to all, no one possessing anything private. He himself, indeed, who most approved of this communion, made use of common possessions in the most just manner; but disciples who changed their minds was given back his original contribution, with an addition, and left. Thus Pythagoras established justice in the best possible manner, beginning at its very first principle.

In the next place, justice is introduced by association with other people, while injustice is produced by unsociabitlity and neglect of other people. Wishing therefore to spread this sociability as far as possibility among men, he ordered his disciples to extend it to the most kindred animal races, considering these as their intimates and friends, which would forbid injuring, slaying or eating any of them. He who recognizes the community of elements and life between men and animals will in much greater degree establish fellowship with those who share a kindred and rational soul. This also shows that Pythagoras promoted justice begin-

ning from its vry root principle. Since lack of
money often compels men sometimes to act contrary to justice, he tried to avoid this by pract-
ising such economy that his necessarie expenses
might be liberal, and yet retain a just suffi-
ciency. For as cities are only magnified households, so the arrangement of domestic concerns is
is the principle of all good order in cities. For
instance, it was said hhat he himself was the
heir to the property of Alcaeus, who died after
completing an embassy to the Lacedemonians; but
that in spite of this Pythagoras was admired for
his economy no less than for his philosophy. Also
when he married he so educated the daughter that
was born to him, and who afterwards married the
Crotonian Meno, that while unmarried she was a
choir-leader, while as wife she held the first
place among those who worshipped at altars. It
is also said that the Metapontines preserved
Pythagoras's memory by turning his house into a
temple of Ceres, and the street in which he lived
into a museum.

Because injustice also frequently results
from insolence, luxury, and lawlessness, he daily
exhorted his disciples to support the laws, and
shun lawlessness. He considered luxury the first
evil that usually glides into houses and cities;
the second insolence, the third destruction.
luxury therefore should by all possible means
be ecluded and expelled; and that from birth
nen should be accustomed live temperately, and
in a manly manner. He also added the necessity
of purification from bad language, whether it
be piteous, or provocative, reviling, insolent
or scurrilous.

Besides this household justice, he added another and most beautiful kind, the legislative,
which both orders what to do and what not to do.
legislative justice is more beautiful than the
judicial kind, resembling medicine which heals
the diseased, but differs in this that it is pre-
ventive, planning the health of the soul from afar.
That is why the best of all legislators

graduated from the school of Pythagoras: first,
Charondas the Catanean, and next Zaleucus and
Timaratus, who legislated for the Locrians. Besides these were Theaetetus and Helicaon, Aristocrates and Phytius, who legislated for the
Rhegini. All these aroused from the citizens honors comparable to those offered to divinities.
For Pythagoras did not act like Heraclitus, who
agreed to write laws for the Ephesians,but also
petulantly added that in those laws he would order the citizens to hang themselves. What laws Py-
Pythagoras endeavored to establish were benevolent and scientific.

Nor need we specially admire those (above-
mentioned professional) legislators. Pythagoras
had a slave by the name of Zamolxis, hailing from
Thrace. After hearing Pythagoras's discourses,
and obtaining his freedom, he returned to the
Getae, and there, as has already been mentioned
at the beginning of this work, exhorted the citizens to fortitude, persuading them that the
soul is immortal. So much so is this that even
at present all the Galatians and Trallians, and
many others of the Barbarians, persuade their
children that the soul cannot be destroyed, but
survives death, so that the latter is not to be
feared, so that (ordinary) danger is to be met
with a firm and manly mind. For instructing the
Getae in these things, and for having written laws
for them, Zamolxis was by them considered as the
greatest of the gods.

Further, Pythagoras conceived that the dominion of the divinities was most efficacious
for establishing justice; and from this principle
he deduced a whole polity, particular laws and
a principle of justice. This his basic theology
was that we should realize God's existence, and
that his disposition towards the human race is
such that he inspects and does not neglect it.
This theology was very useful: for we require
an inspection that we would not be disposed to
resist, such as the inspective government of
the divinity, for if divine nature is of this
nature, it deserves the empire of the universe.

For the Pythagoreans rightly taught that (the
natural) man is an animal naturally insolent,
and changeable in impulse, desire and passions.
He therefore requires an extraordinary inspec-
tionary government of this kind, which may produc
use some chastening and ordering. They therefore
thought that any who recognize their changeable-

ness suould never be forgetful of piety towards

and worship of divinity; ever keeping Him before
the eye of the mind, as watching and inspecting
the conduct of mankind. Every one should pay
heed ,beneath the divine nature, and that of
the genii, to his parents and the laws, and obey
them unfeignedly and faithfullyy In general,
they thought it necessary to believe that there is
is no evil greater than anarchy; since the human re
race is not naturally adapted to salvation with-
wut some guidance.

The Pythagoreans also considered it advisable to adhere to the customs and laws of their
ancestors, even though somewhat inferior to other
regulations. For it is unprofitable and not salutary to evade existing laws, or to be studious of
innovation. Pythagoras, therefore, to evince that
his life was conformable to his doctrines gave
many other specimens of piety to the Gods.

It may be quite suitable to mention one of
these, as example of the rest. I will relate what
Pythagoras said and did relative to the embassy
from Sybaris to Crotona, relative to the return
of the exiles. By order of the ambassadors, some
of his associates had been slain, a part of them,
indeed, by one of the ambassadors himself, while
another one of them was the son of one of those
who had excited the sedition, and had died of
disease. When the Crotonians therefore were
deliberating how they should act in this affair,
Pythagoras told his disciples he was displeased
that the Crotonians should be so much at odds
over the matter, and that in his opinion the
ambassadors should not even be permitted to lead
victims to the altar, let alone drag thence the

suppliant exiles. When the Sybarites came to him
with their complaints, and the man who had slain
some of his disciples with his own hands, was
defending his conduct, Pythagoras declared he
would make no answer to (a murderer). Another
(ambassador) accused him of asserting that he
was Apollo, because when, in the past, some person had asked him about a certain subject, why
the thing was so; and he had retorted, Would he
think it sensible, when Apollo was delivering
oracles to him, to ask Apollo why he did so?
Another one of the ambassadors derided his school,
wherein he taught the return of souls to this
world, saying that, as Pythagoras was about to
descend into Hades, the ambassador would give
Pythagoras an epistle to his father, and begged
him to bring back an answer, when he returned.
Pythagoras responded that he was not about to
descend into the abode of the impious, where he
clearly knew that murderers were punished. As
then the rest of the ambassadors reviled him,
Pythagoras, followed by many people, went to the
sea-shore, and sprinkled himself with water. After
reviling the rest of the ambassadors, one of the
Crotonian counsellors observed that he understood they had defamed Pythagoras, whom not even
a brute would dare to blaspheme, though all animals should again utter the same voice as men,
as prehistoric fables relate.

Pythagoras discovered another method of restraining men from injustice: the fear of judgment.
He knew that this method could be taught, and
that fear was often able to suppress justice.
He asserted therefore that it is much better to
be injured, than to kill a man; for judgment is
dispensed in Hades, where the soul and its essence and the first nature of beings, are accurately appraised.

Desiring to exhibit among human unequal, indefinite and unsymmetrical affairs the equality,
definiteness and symmetry of justice, and to show
how it ought to be exercised, he likened justice
to (a right-angled) triangle, the only one among
geometrical forms, which, though having an in-

finite diversity of adjustments of indeed unequal
parts (the length of the sides), yet has equal
 powers (the square on the hypothenuse is equal
to the squares on the other two sides).
     Since all associations(imply relations with
some other person) and therefore entail justice,
the Pythagoreans declared that there were two
kinds of associations, that differed: the seasonable
able, and the unseasonable, according to age,
merit, familiarity, philanthropy, and so forth.
For instance, the association of a younger
person with an elderly one is useasonable, while
that of two young persons is seasonable. No kind
of anger, threatening or boldness is becoming in
a younger towards an elderly man, all which
unseasonable conduct should be cautiously avoided. So also with respect to merit, for, towards
a man who has arrived at the true dignity of
consummate virtue, neither unrestrained form of
speech, nor any other of the above manners of
conduct is seasonable.
     Not unlike this was what he taughtabout the
relations towards parents and benefactors. He
said that the use of the opportune time was
various. For of those who are angry or enraged,
some are so seasonably, and some unseasonably.
The same distinction obtains with desires,
impulsions and passions, actions, dispositions,
associations and meetings. He further observed that
that to a certain extent, the opportuneness
is to be taught, and that also the unexpected
might be analysed artificially; while none of
the above qualifications obtain when applied
universally, and simply. Nevertheless its results
are very similar to those of opportuneness,
namely elegance, propriety, congruence, and the
like.
     Reminding us that unity is the principle of
the universe, being its principal element, so also
 also is it in science, experiment, and growth.
However two-foldness is most honorable in houses,
cities, camps, and such like organizations. For
in sciences we learn and judge not by any single

hasty glance, but by a thorough examination of
every detail. There is therefore grave danger
of entire misapprehension of things, when the
principle has been mistaken; for while the true
principle remains unknown, no consequent conclusions can be final. The same situation obtains in
things of another kind. Neither a city nor a house
van be well organized unless each has an effective ruler, who governs voluntary servants. For
voluntariness is as necessary with the ruler to
govern, as in the ruled, to obey. So also must
there be a concurrence of will between teacher
and learner; for no satisfactory progress can be
made while there obtains resistance on either
side. Thus he demonstrated the beauty of being
persuaded by rulers, and to be obedient to preceptors.

This was the greatest objective illustration
of this argument. Pherecydes the Syrian had been
his teacher, but now was afflicted with the
morbus pedicularis, Pythagoras therefore went
from Italy to Delos, to nurse him, tending him
until he died, and piously performing whatever
funeral rites were due to his former teacher.
So diligent was he in discharge of his duties
towards those from whom he had received instruction.

Pythagoras insisted strenuously with his
disciples on the fufilment of mutual agreements.
(Here is an illustration). Lysis had once completed his worship in the temple of Juno, and
was leaving as he met in the vestibule with
Euryphamus the Syracusan, one of his fellow-
disciples, who was then entering into the temple.
Euryphamus asked Lysis to wait for him, till he
had finished his worship also. So Lysis sat down
on a stone seat there situate, and waited. Euryphamus went in, finished his worship, but, having become absorbed in some profound considerations, forgot his appointment, and passed out
of the temple by another gate. Lysis however
continued to wait, without leaving his seat,
the remainder of that day, and the following
night, and also the greater part of the next day.

He might have staid there still longer, perhaps,
unless, the following day, in the auditorium, Euryphamus had heard that Lysis's associates were
missing him. Recollecting his appointment, he
hastened to Lysis, relieved him of the engage-
ment, telling him the cause of his forgetfulness
as follows: "Some God produced this oblivion in
me, as a trial of your firmness in keeping your
engagements."

Pythagoras also ordained abstinence from animal food, for many reasons, besides the chief
one that it conduced to peaceableness. Those
who are trained to abominate the slaughter of
animals as iniquitous and unnatural will not
think it much more unlawful to kill a man, or
engage in war. For war promotes slaughter, and
legalizes it, increasing it, and strengthening it.

Pythagoras's maxim "not to touch the balance
above the beam" is in itself an exhortation to
justice, demanding the cultivation of everything
that is just, -- as will be shown when we study
the Pythagorean symbols. In all these particulars,
therefore, Pythagoras paid great attention to
the practice of justice; and to its preachment
to men, both in deeds and words.

CHAPTER XXXI.

TEMPERANCE AND SELF-CONTROL

Temperance is our next topic, cultivated as it was by Pythagoras, and taught to his associates. The common precepts about it have already been deatailed, in which we learned that everything irregular should be cut off with fire and sword. A similar precept is the abstaining from animal food, and also from such likely to produce intemperance, and lulling the vigilance and genuine energies of the reasoning powers. A further step in this direction is the precept to introduce, at a banquet, sumptuous fare, which is to be shortly sent away, and given to the servants, having been exhibited merely to chasten the desires. Another one was that none but courtesans should wear gold, not the free women. Further the practise of taciturnity, and even entire silence, for the purpose of governing the tongue. Next, intensive and continuous puzzling out of the most difficult speculations, for the sake

of which wine, food and sleep would be minimized. Then would come genuine discrediting of notoriety, wealth, and the like; a sincere reverence towards those to whom reverence is due; joined with an unassumed democratic geniality towards one's equals in age, and towards the juniors guidance and counsel, free from envy, and everything similar which is to be deduced from temperance.

The temperance of the Pythagoreans, and how Pythagoras taught this virtue, may be learned from what Hippobotus and Neanthes narrate of Myllias and Timycha, who were Pythagoreans. It seems that Dionysius the tyrant could not obtain the friendship of any one of the Pythagoreans, though he did everything possible to accomplish that purpose; for they had noted, and condemned his monarchical leanings. He therefore sent a troop of thirty soldiers, under the command of Eurymenes the Syracusan, who was the brother of Dion, through (whose) treachery he hoped to take advantage of the Pythagoreans' usual annual migration to catch some of them; for they were in the habit of changing their abode at different seasons of the year, and they selected places suitable to such a migration. Therefore in Phalae, a rugged part of Tarentum, through which the Pythagoreans were scheduled to pass, Eurymenes ...... insidiously concealed his troop; and when the unsuspecting Pythagoreans reached there about noon, the soldiers rushed upon them with shouts, after the manner of robbers. Disturbed and terrified at an attack so unexpected, at the superior number of their enemies, -- the Pythagoreans amounting to no more than ten, -- and being unarmed against regularly equipped soldiery, the Pythagoreans saw that they would inevitably be taken captive, so they decided that their only safety lay in flight, which they did not consider inadmissible to virtue. For they knew that, according to right reason, fortitude is the art of avoiding as well as enduring. So they would have escaped, and their pursuit would have been given up by Eurymenes's soldiers, who were heavily armed, had their flight not led them up against a field sown with beans, which were already flowering. Unwilling to violate their principle not to touch beans, they stood still, and driven to desperation turned, and attacked their pursuers with stones and sticks, and whatever they found to hand, till they had wounded many, and slain some. But (numbers told), and all the Pythagoreans were slain by the s pearmen, as none of them would suffer himself to be taken captive, preferring death, according to the Pythagorean teachings.

As Eurymenes and his soldiers had been sent for the express purpose of taking some of the Pythagoreans alive to Dionysius, they were much crest-fallen; and having thrown the corpses in a common sepulchre, and piled earth thereupon, they turned homewards. But as they were returning they met two of the Pythagoreans who had lagged behind, Myllias the Crotonian, and his Lacedemonian wife Timycha, who had not been able to keep up with the others, being in the sixth month of pregnancy. These therefore the soldiers gladly

made captive, and led to the tyrant with every
precaution, so as to insure their arrival alive.
On learning what had happened, the tyrant was
very much disheartened, and said to the two Pythagoreans, "You shall obtain from me honors of
unusual dignity if you shall be willing to reign
in partnership with me."All his offers, however,
were by Myllias and Timycha rejected. Then said
he, I will release you with a safe-guard if you
will tell me one thing only. On Myllias asking
what he wished to learn, Dionysius replied: "Tell
me only why your companions chose to die rather
than to tread on beans?" But Myllias at once answered, "My companions did indeed prefer death
to treading on beans; but I had rather do that
than tell you the reason." Astonished at this ,
answer, Dionysius ordered him removed forcibly,
and Timycha tortured, for he thought that a pregnant woman, deprived of her husband, would weak-
en before the torments, and easily tell him all
he wanted to know. The heroic woman, however,
with her teeth bit her tongue until it was separated and spat it out at the tyrant, thus demonstrating that the offending member should be
entirely cut off, even if her sexweakness, vanquished by the torments, should be compelled to
disclose something that should be reserved in
silence. Such difficulties did they make to the
admission of outside friendships, even though
they happened to be royal.

Similar to these also were the precepts concerning silence, which tended to the practice of
temperance; for of all continence, the subjugation of the tongue is the most difficult. The same
virtue is illustrated by Pythagoras's persuading
the Crotonians to relinquish all   sac-
rilegious and questionable commerce with courtesans. Moreover Pythagoras restored to temperance a youth who had become wild with amatory
passion, through music. Exhortations against
lascivious insolence promote the same virtue.

Such things were delivered to the Pythagoreans by Pythagoras himself, who was their cause.

They took such care of their bodies that they
remained in the same condition, not being at one
time lean, and at another stout, which changes
they considered anomalous. With respect to their
mind also, they managed to remain uniformly mil-
dly joyful, and not at one time hilarious, and
at another sad, which could be achieved only by
expelling perturbations, despondency or rage.

It was a precept of theirs that no human
casualties ought to be unexpected by the intelligent, expecting everything which it is not in
their power to prevent. If however at any time any
of them fell into a rage, or into despondency,
he would withdraw from his associates' company,
and seeking solitude, endeavor to digest and
heal the passion.

Of the Pythagoreans it is also reported that
none of them punished a servant or admonished a
free man during anger; but waited until he had
recovered his wonted serenity. They used an especial word, **paidartan**, to signify such (self-con-
trolled) rebukes, effecting this calming by silence and quiet. So Spintharus relates of Archytas
the Tarentine that on returning after a certain
time from the war against the Messenians waged by
the Tarentines, to inspect some land belonging
to him, and finding that the bailiff and the other
servants had not properly cultivated it, greatly
neglecting it, he became enraged, and was so furious that he told his servants that it was well
for them that he was **angry**, for otherwise they
would not have escaped the punishment due to so
great an offence. A similar anecdote is related
of Clinias, according to Spintharus; for he also
was wont to defer all admonitions and punishments
until his mind was restored to tranquility.

Of the Pythagoreans it is further related that
they restrained themselves from all lamentation,
weeping, and the like; and that neither **gain**,
desire, anger or ambition, or anything of the like,
ever became the cause of dissension among them;
all Pythagoreans being disposed towards each other
as parents towards their offspring.

Another beautiful trait of theirs was that they gave credit to Pythagoras for everything, naming it after him, not claiming the glory of their own inventions, excpt very rarely. Few are there who acknowledged their own works.

Admirable too is the careful secrecy with which they preserved the mystery of their writings. For during so many centuries , prior to the times of Philolaus, none of the Pythagorean commentaries appeared publicly. Philolaus first published those three books celebrated books which, at the request of Plato, Dion of Syracuse is said to have bought for a hundred minae. For Philolaus had been overtaken by sudden severe poverty, and he capitalized the writings of which he was partaker through his alliance with the Pythagoreans.

As to the value of opinion, such were their views: A stupid man should defer to the opinion of any one, especially to that of the crowds. Only a very few are qualified to apprehend and opine rightly; for evidently this is limited to the intelligent, who are very few. To the crowds, such a qualification of course does not extend. But to despise the opinion of every one is also stupid; for such a person will remain unlearned and incorrigible. The unscientific should study that of which he is ignorant, or lacks scientific knowledge. A learner should also defer to the opinion of the scientific, and is able to teach. Generally, youths who wish to be saved should attend to the opinions of their elders, or of those who have lived well.

During the course of human life there are certain ages, by them called endedasmenae, which cannot be connected by the power of any chance person. Unless a man from his very birth is trained in a beautiful and upright manner, these ages antagonize each other. A well educated child, formed to temperance and fortitude, should devote a great part of his education to the stage of adolescence. Similarly, when the adolescent is trained to temperance and fortitude, he should focus his education on the next age of manhood. Nothing could be more absurd than the way in which the general public

treats this subject. They fancy that boys should
be orderly and temperate, abstaining from everything troublesome or indecorous; but as soon as
they have arrived at the age of adolescence, they
may do anything they please. In this age, therefore, there is a combination of both kinds of errors, puerile and virile. To speak plainly, they
avoid anything that demands diligence and good
order, while following anything that has the appearance of sport, intemperance and petulance,
being familiar only with boyish affairs. Their
desires should be developed from the boyish stage
stage into the next one. In the meanwhile ambition and the rest of the more serious and turbulent inclinations and desires of the virile
age prematurely invade adolescence; wherefore
this adolescence demands the greatest care. I

In general, no man ought to be allowed to
do whatever he pleases; but there is always
need of a certain inspection, or legal and cul-
tured government, to which each of the citizens
is responsible. For animals, when left to themselves, and neglected, rapidly degenerate into
vice and depravity.

The Pythagoreans (who did not approve of
men being intemperate,)would often compel answers from, and puzzle (such intemperate people)
by asking them why boys are generally trained
to take food in an orderly and moderate manner,
being compelled to learn that order and decency
are beautiful,and their contraries, disorder and
intemperance base, while drunkards and gorman-
izers are held in great disgrace. For if none
of these (temperate) habits are to be continued
on into the virile age, to accustom us, as boys,
to such (temperate) habits, was useless. The
same argument holds good in respect to other
good habits to which children are trained. Such
a reversal of training is not seen in the case
of the education of other lower animals. From
the very first a whelp and a colt are trained
to, and learn those tricks which they are to
exercise when arrived at maturity. (The more
liberal standard for men in the matter of morals

is therefore not sustained by the common sense
that trains children to temperance).

The Pythagoreans are generally reported to
have exhorted not only their intimates, but also
whomsoever they happened to meet, to avoid pleasure as a danger demanding the utmost caution.
More than anything else does this passion deceive
us, and mislead us into error. They contended
that it was wiser never to do anything whose end
was pleasure, whose results are usually shameful
and harmful. They asserted we should adopt as
end the beautiful and fair, and do our duty. Only
secondarily should we consider the useful and
advantageous. In these matters there is no need
to consider considerations of chance.

Of desire, the Pythagoreans said: That desire itself is a certain tendency, impulse and
appetite of the soul, wishing to be filled with
something, or to enjoy the presence of something,
or to be disposed according to some sense-en-
joyment. There are also contrary desires, of evacuation and repulsion, and to terminate some
sensation. This passion is manifold, and is almost the most Protean of human experiences.
However, many human desires are artificially acquired, and self-prepared. That is why this passion demands the utmost care and watchfulness,
and physical exercise that is more than casual.
That when the body is empty it should desire
food is no more than natural; and then it is
just as natural that when it is full it should
desire evacuation. appropriate evacuation. But
to desire superfluous food, or luxurious garment.
or coverlets, or residences, is artificial. The
Pythagoreans applied this argument also to furn-
iture, dishes, servants and cattle raised for
butchering. Besides, human passions are never
permanent, but are ever changing, even to infinity. That is why education of the youth should
begin at the earliest moment possible, that their
aspirations may be directed towards ends that
are proper, avoiding those that are vain and
unnecessary, so as to be undisturbed by, and
remain pure from such undesirable passions;

and may despise such as are objects of contempt,
because   subjected to changeable desires. Yet
it must be observed that senseless, harmful,
superfluous and insolent desires subsist in the
souls of such individuals who are the most powerful; for there is nothing so absurd that the
soul of such boys, men and women would not lead
them to perform.

Indeed, the variety of food eaten is beyond
description. The kinds of fruits and roots which
the human race eats is nothing less than infinite. The kinds of flesh eaten are innumerable;
there is no terrestrial, aerial, or aquatic animal which has not been partaken of. Besides, in
the preparation of these, the contrivances used
are innumerable, and they are seasoned with manifold mixtures of juices. Hence, according to
the motions of the human soul, it is no more
than natural that the human race should be so
various as to be actually insane; for each
kind of food that is introduced into the human body becomes the cause of a certain peculiar disposition.

(Quantity) is as important as quality); for
sometimes a slight change in quantity produces
a great change in quality, as with wine. First
making men more cheerful, later it undermines
morals and sanity. This difference is generally
ignored in things in which the result is not so
pronounced, although everything eaten is the cause
of a certain peculiar disposition. Hence it requires great wisdom to know and perceive what quality and quantity of food to eat. This science,
first unfolded by Apollo and Phaon, was later
developed by Aesculapius and his followers.

About propagation, the Pythagoreans taught
as follows. First, they prevented untimely
birth. Not even among plants or animals is prem-
aturity good. To produce good fruit there is
need of maturation for a certain time to give
strong and perfect bodies to fruits and seeds.
Boys and girls should therefore be trained to
work and exercises, with endurance, and that

they should eat foods adapted to a life of labor
and temperance, with endurance. There are many
things in human life which it is better to learn
at a late period in life, and this sex-life is
one of them. It is therefore advisable that a
boy should be educated so as not to begin sex-
connection before the twentieth year, and even
then rarely. This will take place if he holds
high ideals of a good habit for the body. Body-
hygiene and intemperance are not likely to subsist in the same individual. The Pythagoreans
praised the earlier Greek laws forbidding intercourse with a woman who is a mother, daughter or sister in a temple or other public place.
It is advisable that there be many impediments
to the practice of this energy. The Pythagoreans
forbad entirely intercourse that was unnatural,
or resulting from wanton insolence, allowing only the natural, the temperate, which occur in
the course of chaste and recognized procreation
of children.

Parents should make circumstancial provision
for their offspring. The first precaution is a
healthful and temperate life, not unseasonably
filling himself with food, nor using foods which
create bad body-habits, above all avoiding intoxication. The Pythagoreans thought that an
evil, discordant, rrouble-making character produced depraved sperma. They insisted that none
but an indolent or inponsiderate person would
attempt to produce an animal , and introduce
it to existence, without most diligently pro-
viding for it a pleasing and even elegant ingress
into this world. Lovers of dogs pay the utmost
possible attention to the breeding of their
puppies, knowing that goodness of the offspring
depends on goodness of parents, at the right seas-
season, and in proper surroundings. Lovers of bir
birds pay no less attention to the matter; pro-
creators of generous animals therefore should by
all possible means manage that their efforts be
fruitful. It is therefore absurd for men to pay
no attention to their own offspring, begetting

CASUALLY AND CARELESSLY, and after birth, feed
and educate them negligently. This is the most
powerful and manifest cause of the vice and depravity of the greater part of mankind, for the
generality undertake procreation on impulse, like
beasts.

Such were the Pythagoreans' teachings about temp
temperance, which they defended by word and practised in deed. They had originally received them from
from Pythagoras himself, as if they had been oracles delivered by the Pythian Apollo himself.

CHAPTER XXXII

FORTITUDE

Fortitude, the subject of this chapter, has already been illustrated, by the heroism of Timycha, and those Pythagoreans who preferred death to transgression of Pythagoras's prohibition to touch beans, and other instances. Pythagoras himself showed it in the generous deeds he performed when travelling everywhere alone, undergoing heart-breaking labors and serious dangers, and in choosing to leave his country and living among strangers. Likewise when he dissolved tyrannies, ordered confused commonwealths, and emancipated cities. He ended illegalities, and impeded the activities of insolent and tyrannical men. As a leader, he showed himself benignant to the just and mild, but expelled rough and licentious men from his society, refusing even to answer them, resisting them with all his might, although he assisted the former.

Of these courageous deeds, as well as of many upright actions, many instances could be adduced; but the greatest of these is the prevailing freedom of speech he employed towards the tyrant Phalaris, the most cruel of them, who detained him in captivity. A Hyperborean sage named Abaris visited him, to converse with him on many topics, especially sacred ones, respecting statues and worship, the divine Providence, natures terrestrial and celestial, and the like. Pythagoras, under divine inspiration, answered him boldly, sincerely and persuasively, so that he converted all listeners. This roused Phalaris's anger against Abaris, for praising Pythagoras, and increased the tyrant's resentment against Pythagoras. Phalaris swore proudly as was his wont, and uttered blasphemies against the Gods themselves. Abaris however was grateful to him, and learned from him that all things are suspended from, and governed by the heavens; which he proved

from many considerations, but especially from the
potency of sacred rites. For teaching him these
things, so far was Abaris from thinking Pythagoras an enchanter, that his reverence for him increased till he considered him a God. Phalaris
tried to counteract this by discrediting divination, and publicly denying there was any efficacy of the sacraments performed in sacred rites.
Abaris, however, guided the controversy towards
such things as are granted by all men, seeking to
persuade him of the existence of a divine providence, from circumstances that lie above human
influence, such as immense wars, incurable diseases, the decay of fruits, incursions of pestilence, or the like, which are hard to endure, and
are deplorable, arising from the beneficent
(purifying) energy of the powers celestial and
divine.

Shamelessly and boldly Phalaris opposed
all this. Then Pythagoras, suspecting that Phalaris intended to put him to death, but knowing
he was not destined to die through Phalaris,
retorted with great freedom of speech. Looking
at Abaris, he said that from the heavens to
to aerial and terrestrial beings there was a
certain descending communication, Then from instances generally known he showed that all things
follow the heavens. Then he demonstrated the
existence of an indisputable power ßß freedom
of will, in thesoul; proceeding further amply to
discuss the perfect energy of reason and intellect. With his (usual) freedom of will he even
(dared to) discuss tyranny, and all the prerogatives of fortune, concerning injustice and human avarice, solidly teaching that all these are
of no value. Further, he gave Phalaris a divine
admonition concerning the most excellent life,
earnestly comparing it with the most depraved.
He likewise clearly unfolded the manner oß subsistence of the soul, its powers and passions;
and, what was the most beautiful of all, demonstrated to him that the Gods are not the authors
of evils, and that diseases and bodily calamities
are the results of intemperance, at the same time

finding fault with the poets and mythologists
for the unadvisedness of many of their fables.
Then he directly confuted Phalaris, and admonished him, experimentally demonstrating to him
the power and magnitude of heaven, and by many
arguments demonstrated to him that reason dict-
ates that punishments should be legal. He demonstrated to him the difference between men and
other animals, scientifically demonstrating the
difference between internal and external speech.
Then he expounded the nature of intellect, and
the knowledge that is derived therefrom; with
its ethical corollaries. Then he discoursed about
the most beneficial of useful things, adding the
mildest possible implied admonitions, adding
prohibitions of what ought not to be done. Most
important of all, he unfolded to him the dis-
tion between the productions of fate and intellect, and the difference between the results of
destiny and fate. Then he reasoned about the
divinities, and the immortality of the soul.

All this, really, belongs to some other
chapter, the present one's topic being the development of fortitude. For if, when situated in
the midst of the most dreadful circumstances
Pythagoras philosophised with firmness of decision, if on all sides he resisted fortune, and
repelled it, enduring its attacks strenuously,
if he employed the greatest boldness of speech
towards him who threatened his life, it must be
evident that he entirely despised those things
generally considered dreadful, rating them as
unworthy of attention. If also he despised exec-
ution, when this appeared imminent, and was not
moved by its imminence, it is evident that he was
was perfectly free from the fear of death, (and
all possible torments).

But he did something still more generous, effecting the dissolution of the tyranny,, restraining
the tyrant when he was about to bring the most
deplorable calamities on mankind, and liberating
Sicily from the most cruel and imperious power.
That it was Pythagoras who accomplished this, is
evident from the oracles of Apollo, which had

predicted that the dominion of Philaris would come
to an end when his subjects would become better
men, and cooperate; which also happened through
the presence of Pythagoras, and his imparting to
them instruction and good principles. The best
proof of this may be found in the timen when it .
happened. For on the very day that Phalaris condemned Pythagoras and Abaris to death, he himself
was by stratagem slain.

Another argument for the truth of this are
the adventures of Epimenides. He was a disciple of
Pythagoras; and when certain persons planned to a
destroy him, he invoked the Furies and the avenging
ing divinities, and thereby caused those who had
attempted his life to destroy each other. In the unic
same way Pythagoras, who assisted mankind, imitating both the manner and fortitude of Hercules,
for the benefit of men punished and occasioned the
death of him who had behaved insolently and in a
disorderly manner towards others; and this through
the very oracles of Apollo, in the class of which
divinity both he and Epimenides had naturally
since birth belonged. This admirable and strenuous deed was the effect of his fortitude.

We shall present another example of preservation of lawful opinion; for following it out, he
did what to him seemed just and dictated by right
reason, without permitting himself to be diverted
from his intention by pleasure, labor, passion or
danger. His disciples also preferred death to
transgression of any precept of his. They preserved
their manners unchanged under the most varying
fortunes. Being involved in a myriad calamities
could not cause them to deviate from his rules.
They never ceased exhorting each other to support
the laws, to oppose lawlessness, from birth to
train themselves to a life of temperance and fortitude, so as to restrain and oppose luxury. They
also used certain original melodies as remedies
against the passions of the soul; against lamenta
tion and despondency, which Pythagoras had invented, as affording the greatest relief in these
maladies. Other melodies they employed against
anger and rage, through which they could in-

crease or diminish those passions, till they reduced them to moderation, and compatibility with
fortitude. The thought which afforded them the
greatest support in generous endurance was the
conviction that no human casualty should be unexpected by men of intellect, but that they must
resign themselves to all vicissitudes beyond
human control.

Moreover, whenever overwhelmed by grief or
anger, they immediately forsook the company of
their associates, and in solitude endeavored
to digest and heal the oppressing passion. They
took it for granted that studies and disciplines
implied labor, and that they must expect severe
tests of different kinds, and be restrained
and punished even by fire and sword, so as to
exercise innate intemperance and greediness;
for which purpose no labor or endurance should
be spared. Further to accomplish this, they un-
selfishly abstained from animal food, and also
some other kinds. This also was the cause of their
slowing   .. of speech and complete silence, as
means to the entire subjugation of the tongue,
which demanded year-long exercise of fortitude.
In addition, their strenuous and assiduous investigation and resolution of the most difficult theorems, their abstinence from wine, food
and sleep, and their contempt of wealth and glory.
Thus by many different means they trained themselves to fortitude.

But this is not all. They restrained themselves from pamentations and tears. They abstained from entreaty, supplication, and adulation, as
effeminate and abject.(or, humble). To the same
practice of fortitude must be referred their
peculiarity of absolute reserve among their
arcana of the principal principles of their discipline, preserving them from being divulged to
strangers, committing them unwritten to memory,
and transmitting them orally to their successors
as if they were the mysteries of the Gods. That
is why nothing worth mentioning of their philosophy was ever made public, and though it had
been taught and learned for a long while, it was

not known beyond their walls. Those outside, whom
I might call the profane, sometimes happened to
be present; and under such circumstances the Pythagoreans would communicate only obscurely, through
symbols, a vestige of which is retained by the
celebrated precepts still in circulation, such as,
Fire should not be poked with a sword, and other
like ones, which, taken literally, resemble old-
wives' tales; but which, when properly unfolded,
are to the recipients admirable and venerable.

That precept which, of all others, was of the
greatest efficacy in the achievement of fortitude
is that one which helps defend and liberate from
the life-long bonds that reatain the intellect
in captivity, and without which no one can perceive or learn anything rational or genuine,
whatever be the sense in activity. They said,
"Tis mind that sees all things, and hears
                     them all;
All else is deaf and blind."
The next most efficacious precept   is that one
which exhorts excessively to be studious of purifying the intellect, and by various methods
adapting it through mathematical disciplines to
receive something divinely beneficial, so as
neither to fear a separation from the body, nor,
when directed towards incorporeal natures, through
their most refulgent splendor to be compelled to
turn away the eyes, nor to be converted to those
passions which fastena and even nail the soul to
the body, and makes her rebellious to all those
passions which are the progeny of procreation,
degrading her to a lower level. The training of
ascent through all these is the study of the most
perfect fortitude. Such are important instances
of the fortitude of Pythagoras and his followers.

CHAPTER XXXIII

UNIVERSAL FRIENDSHIP

Friendship of all things towards all was most clearly enforced by Pythagoras. God's friendship towards men he explained through piety and scientific cultivation; but that of teachings towards each other, and generally of the soul to the body, of the rational towards the irrational part, through philosophy and its teachings; . . . . That of men towards each other, and of citizens, he justifie through proper legislation; that of strangers, through the common possession of a body; that between man and wife, children, brothers or kind red, through the unperverted ties of nature. In short, he taught the friendship of all for all; and still further, of certain animals, through justice, and common physical experiences. But the pacification and conciliation of the body,he which is mortal by itself, and of its latent immortal powers, he enforced through health, and a temperate diet suitable thereto, in imitation of the ever-healthy condition of the mundane elements.

In all these, Pythagoras is recognized as the inventor and summarizer of them in a single name, that of friendship. So admirable was his friendship to his associates, that even now when people are extremely benevolent mutually people call them Pythagoreans. We should therefore narrate Pythagoras's discipline thereto related, and the precepts he taught his disciples.

Pythagoreans therefore on the effacing of all rivalry and contention from true friendship, if not from all friendship; at least from parental friendship, and generally from all gratitude towards seniors and benefactors. To strive or contend with such, out of anger or some other pas sion, is not the way to preserve existing friend ship. Scars and ulcers in friendship should be the least possible; and this will be the case if those that are friends know how to subdue their

anger. If indeed both of them know this, or rather
the younger of the two, and who ranks in some one
of the above-mentioned orders, (their friendship will be the more easily preserved). They
also taught that corrections and admonitions,
which they called "paedartases" should take place
from the elder to the younger, and with much suavity and caution; and likewise thatmuch careful
and considerate attention should be manifested
in admonitions.For thus they will be persuasive
and helpful. They also said that confidence shoul
should never be separated from   friendship,
whether in earnest, or in jest. Existing friendship cannot survive, when once falsehood insin-
uates itself into the habits of professed friends.
According to them, friendship should not be aban-
doned on account of misfortune, or any other
human vicissitude; the only permissible rejection of friend or friendship is the result of
great and incorrigible vice. Hatred should not
be entertained voluntarily against those who are
not perfectly bad, but when once formed, it
should be strenuously and firmly maintained,
unless its object should change his morals, so
as to become a better man. Hostility should not
consist in words, but in deeds. War is commendable and legitimate, when conducted in a manly
manner.

No one should ever permit himself to become
the cause of contention, and we should se far as
possible avoid its source. In a friendship which
is intended to be pure, the greater part of the
things pertaining to it should be definite and
legitimate. These should be properly distinguished,
an. not be casual; and moreover our conversation
should never grow casual or negligent, but remain orderly, modest and benevolent. So also with
the remaining passions and dispositions.

We should not decline foreign friendships carelessly, but accept and guard them with the greatest care;

That the Pythagoreans preserved friendship
towards each other for many ages may be inferred

Here is a specimen of what it contained:
(This next paragraph is misplaced, but is put
here as more suitable here than where it is
in the text, in front of the last one).

None of the Pythagoreans called Pythagoras
by his name. While alive, they referred to him
as the divine one; after his death, as that man,
just as Homer makes Eumaeus refer to Ulysses
thus:

Though absent he may be, O guest, I fear
To name him; so great is my love and care."

Such were some of his precepts: They were to
get up before sunrise, and never to wear a ring
on which the image of God was engraved, lest that
image be defiled by being worn at funerals, or
other impure place. They were to adore the rising
sun. Pythagoras ordered them never to do anything without previous deliberation and discussion; in the morning forming a plan of what was
to be done later, and at night to review the
day's actions, which served the double purpose
of strengthening the memory, and considering
their conduct. If any of their associates appointed them to meet them at some particular place and
time, they should stay there till he came, regard-
less of the length of time, for Pythagoreans shoul
should not speak carelessly, but remember what
was said, and regard order and method. At death
they were not to blaspheme, but to die uttering
propitious words, such as are used by those who
sail out of the port into the Adriatic Sea.

Friends are to be venerated in the same manner
as Gods; but others are to be treated as brutes.
This very sentiment is ascribed to Pythagoreans
themselves, but in verse-form; such as,
" Like blessed Gods, his friends he e'er revered,
But reckoned others as of no account." Pythagora
considered that Homer deserved to be praised for
calling a king the shepherd of the people, which
implied approval of aristocracy, in which the
rulers are few, while the implication is that
the rest of men are like cattle. Enmity was required to beans, because they were used in vot-
ing; inasmuch as the Pythagoreans selected of-

ficeholders by appointment. To rule should be an
object of desire, for it is better to be a bull
for one day only, than for all one's life to be
an ox. While other states' constitutions might
be laudable, yet it would be advisable to use
only that which is known to oneself.

In short, Ninon showed that their philosophy
was a conspiracy against democracy; and advised
the people not even to listen to the defendants,
considering that they would never have been admitted into the assembly if the Pythagoreans' council had had to depend for admission on the session
of a thousand men; that they should not allow
speech to those who, had used their utmost power
to prevent speech by others. The people must remember that when they raised their right hands to
vote, or even counted their votes, this their
right hand was constructively rejected by the
Pythagoreans, who were aristocrats. It was also
disgraceful that the Crotonian masses who had
conquered thirty myriads of men at the river
Tracis should be outweighed by a thousandth part
of the same number through sedition in the city
itself. Through these calumnies Ninon so exas-
perated his hearers that a few days after a great
multitude assembled intending to attack the Pythagoreans as they were sacrificing to the Muses
in a hose near the temple of Apollo. Foreseeing
this, the Pythagoreans fled to an inn, while Democedes with the youths retired to Plataea. The
partisans of the new constitution decreed an accusation against Democedes of inciting the youths
to capture power, putting a price of thirty talents on his head, dead or alive. A battle ensued,
and and the victor Theages was given the thirty
talents promised by the city. The city's evils
were spread to the whole region, and the exiles
were arrested even in Tarentum, Metapontum and
Caulonia, The envoys from these cities that came
to Crotona to get the charges were, according
to the Crotonian record, bribed, with the result
that the exiles were condemned as guilty, and
driven out further. The Crotonians then expelled
from the city all who were dissatisfied with

symbol, found out who had placed the tablet
there, and having also investigated every par
ticular, paid the inn-keeper a very much greater
sum than he had disbursed.

It is also related that Clinias the Tarent
when he learned that the Cyronaean Prorus , wh
was a zealous Pythagorean, was in danger of los
ing all his property, sailed to Cyrene, and al
having collected a sum of money, restored the a
fairs of Prorus to a better condition, though
thereby he sensibly diminished his own estate,
and risked the peril of the sea-voyage.

Similarly, Thestor Posidoniates, having from
more report heard that the Pythagorean Thymarid a
Parius had fallen into poverty, from great wealth
into abject poverty, is said to have sailed to
Paros, and after having collected a large sum o
money, and reinstated Thymaridas in affluence.
These are beautiful instances of friendship.

But much more admirable than the above exampl
were the Pythagoreans' teachings respecting the
communion of divine goods, the agreement of intellect, and their doctrines about the divine
soul,   They were ever exhorting each other not to tear apart the divine
soul within them. The significance of their frie
ship   both in words and deeds was
effort to achieve a certain divine union, (or
union with the divinity), or communion of intelle
with the divine soul. Better than this, either i
what is uttered in words , or performed by does
it is not possible to find. For I am of opinion
that in this all the goods of friendship are un
ited. In this, as a climax, we have collected
all the blessings of Pythagorean friendship; the
is nothing left to say.

CHAPTER XXXIV

NON-HEREDITARY SECRECY

Having thus, according to plan, discussed
Pythagoras and Pythagoreanism, we may be interested in scattered points which do not fall under any of the former topics.

(First, as to language). It is said that each
Greek novice was ordered to use his native language, as they did not approve of the use of a foreign language. Foreigners joined the Pythagoreans:
Messenians, Lucani, Picentini, and Romans. Metrodorus, the son of Thyrsus, who was the father
of Epicharmus, who specialised in medicine, in
explaining his father's writings to his brother,
says that Epicharmus, and prior to him Pythagoras,
conceived that the best dialect, and the most
musical, was the Doric. The Ionic and Aeolic
remind of chromatic progression, which however
is still more evident in the Attic. The Doric,
consisting of pronounced letters, is enharmonic.

The myths also bear witness to the antiquity
of this dialect. Nereus was said to have married
Doris, the daughter of Ocean; by whom he had fif-
ty daughters, one of whom was the mother of Ach-
illes. Metrodorus also says that some insist that
Helen was the offspring of Deucalion, who was the
son of Prometheus and Pyrrha the daughter of Epi-
metheus; and from him descended Dorus and Aeolus.
Further he observes that from the Babylonian
sacred rites he had learned that Helen was the
offspring of Jupiter, and that the sons of Hellen
were Dorus, Xuthus, and Aeolus; with which Herodotus also agrees. Accuracy in particulars so
ancient is difficult for moderns, to enable them
to decide which of the accounts is most trustworthy. But either of them claim that the Doric
dialect is the most ancient, that the Aeolic,
whose name derives it from Aeolus, is the next
in age, and that the third is the Ionic, derived
from Ion, the son of Xuthus. Fourth is the Attic,
named from Creusa, the daughter of Erechtheus,
and is three generations younger than the others;

as it existed about the time of the Thracians,
and the rape of Orithyia, as is evident from
the testimony of most histories. The Doric di
lect was also used by the most ancient of the
poets, Orpheus.......... (repetition).

The Pythagoreans objected to those who o
fered disciplines for sale, who open their so
like the gates of an inn to every man that a
proaches them; and who, if they do not thus
buyers, diffuse themselves through cities, a
in short, hire gymnasia, and require a reward
from young men for those things that are with
price. Pythagoras indeed hid the meaning of m
that was said by him, in order that those who
were genuinely instructed might clearly be pa
takers of it; but that others, as Homer says
Tantalus, might be pained in the midst of wha
they heard, in consequence of receiving no de
light therefrom.

The Pythagoreans thought that those who tea
for the sake of reward, that they show themsel
worse than sculptors, or artists who perform t
work sitting. For these, when someone orders t
to make a statue of Hermes, search for wood suit
to receive the proper form; while those pretend
that they can readily produce the works of vir
from every nature.

The Pythagoreans likewise said that it is
more necessary to pay attention to philosophy,
than to parents or agriculture; for no doubt it
is owing to the latter that we live, but philos
ophers and preceptors are the causes of our liv
ing well, and becoming wise, on discovering the
right mode of discipline and instruction.

Nor did they think fit either to speak or t
write in such a way such that their conceptions
might be obvious to the first comer; for the ve
first thing Pythagoras is said to have taught i
that, being purified from all intemperance, his
disciples should preserve the doctrines they ha
heard in silence. It is accordingly reported th
he who first divulged the theory of commensurab
and incommensurable quantities to those unworthy
to receive it, was by the Pythagoreans so hated

that they not only expelled him from their common
association, and from living with him, but also
for him constructed a tomb, as for one who had
migrated from the human into another life. It is
also reported that the Divine Power was so indignant with those who divulged the teachings of Pythagoras, that he perished at sea, as an impious
person who divulged the method of inscribing in
a sphere the dodecahedron, on of the five so-called
solid figures, the composition of
the icostagonus. But according to others, this
is what happened to him who revealed the doctrine
of irrational and incommensurable quantities.

All Pyhtagoric discipline was symbolic, resembling riddles and puzzles, and consisting of maxims,
in the style of the ancients. Likewise the truly divine
Pythian oracles seem to be somewhat difficult of
understanding and explanation; to those who carelessly receive the answers given. These are the
indications about Pythagoras and the Pythagoreans
collected from tradition.

CHAPTER XXXV

ATTACK ON PYTHAGOREANISM

There were however certain persons who were
hostile to the Pythagoreans, and who rose against
them. That stratagems were employed to destroy
them, during Pythagoras's absence, is universally
acknowledged; but the historians differ in their
account of the journey which he then undertook.
Some say that he went to Pherecydes the Syrian,
and others, to Metapontum. Many causes of the
stratagems are assigned. One of them, which is also
said to have originated from the men called Cylonians, is as follows: Cylon of Crotona was one of
the most prominent citizens, in birth, renown
and wealth; but in manners he was severe, turbulent, violent, tyrannical. His greatest desire
was to become partaker of the Pythagoric life,
and he made application to Pythagoras who was now
advanced in age, but was rejected for the above
causes. Conseqently he and his friends became
violent enemies of the brotherhood. Cylon's ambition was so vehement and immoderate that with
his associates he persecuted the very last of the
Pythagoreans. That is why Pythagoras moved to
Metapontum, where he closed his existence.

Those who were called Cylonians continued to
plot against the Pythagoreans, and to exhibit the most virulent malevolence. Nevertheless,
for a time this enmity was subdued by the Pythagoreans' probity, and also by the vote of the citizens, who entrusted the whole of the city affairs
to their management.

At length, however, the Cylonians became
so hostile to "the men," as they were called,
that they set fire to Milo's residence, where
were assembled all the Pythagoreans, holding a
council of war. All were burnt, except two, Archippus and Lysis, who escaped through their
bodily vigor. As no public notice was taken of
this calamity, the Pythagoreans ceased to pay
any further attention to public affairs, — which
was due to two causes: the cities' negligence,

and through the loss of those men most qualified
to govern.

Both of the saved Pythagoreans were Tarentines, and Archippus returned home. Lysis, res-
enting the public neglect, went into Greece, residing in the Achaian Peloponnesus. Stimulated by
an ardent desire, he migrated to Thebes, where
he had as disciple Epaminondas, who spoke of his
teacher as his father. There Lysis died.

Except Archytas of Tarentum, the rest of the
Pythagoreans departed from Italy, and dwelt together in Rhegium. The most celebrated were Phanto,
Echecrates, Polymnastus, and Diocrates, who were
Phliasians; and Xenophilus Chalcidensis of Thrace.
But in course of time, as the administration of
public affairs went from bad to worse, these
Pythagoreans nevertheless preserved their pristine manners and disciplines; yet soon the sect
began to fail, till they nobly perished. This is
the account by Aristoxenus.

Nichomachus agrees with Arisoxenus, except
that he dates the plot against the Pythagoreans during Pythagoras's journey to Delos, to
nurse his preceptor Pherecydes the Syrian, who
was then afflicted with the morbus pedicularis,
and after his death performed the funeral rites.
Then those who had been rejected by the Pythagoreans , and to whom monuments had been raised,
as if they were dead, attacked them, and committed them all to the flames. Afterwards they were
they were overwhelmed by the Italians with stones,
and thrown out of the house unburied. Then science
died in the breasts of its possessors, having by
them been preserved as something mystic and
incommunicable. Only such things as were difficult to be understood, and which were not
expounded, were preserved in the memory of those
who were outside the sect, except a few things,
which certain Pythagoreans, who at that time happened to be in foreign lands, presrrved as
sparks of science very obscure, and of difficult
investigation. These men beibg solitary, and dejected at this calamity, were scattered in differ-

ent places, retaining no longer any public influence. They lived alone in solitary places,
wherever they found any; each preferred association with himself to that with any other person.

Fearing however lest the name of philosophy
should be entirely exterminated from among mankind, and that they should, on this account, in-
cur the indignation of the Gods, by suffering so
great a gift of theirs to perish, they made an
arrangement of certain commentaries and symbols
gathered the writings of the more ancient Pythagoreans, and of such things as they remembered.
These relics each left at his death to his son,
or daughter, or wife, with a strict injunction
not to alienate from the family. This was carried
out for some time, and the relics were transmitted in succession to their posterity.

Since Apollonius dissents in a certain place
regarding these particulars, and adds many things
that we have not mentioned, we must record his
account of the plot against the Pythagoreans.
He says that from childhood Pythagoras had aroused envy. So long as he conversed with all that
came to him, he was pleasing to all; but when he
restricted his intercourse to his disciples the
general people's good opinion of him was altered. They did indeed permit him to pay more attention to strangers than to themselves; but they
were indignant at his preferring some of their
fellow-citizens before others; and they suspected that his disciples assembled with intentions
hostile to themselves. In the next place, as the
young men that were indignant with him were of
high rank, and surpassed others in wealth, and,
when they arrived at the proper age, not only
held the first honors in their own families,
but also managed the affairs of the city in
common, they, being more than three hundred in
number, formed a large body, so that there remained but a small part of the city which was not
devoted to their hbits and pursuits.

Moreover so long as the Crotonians confined
themselves to their own country, and Pythagoras
dwelt among them, the original form of government
continued; but the people had changed, and they
were no longer satisfied with it, and were therefore seeking a pretext for a change. When they
captured Sybaris, and the land was not divided by
lot, according to the desire of the multitude,
and Pythagoras gone, this veiled hatred against
the Pythagoreans burst forth, and the populace
forsook them.

The leaders of this dissension were those
that were nearest to the Pythagoreans, both by
kindred and intercourse. These leaders, as well
as the common folk were offended by the Pythagoreans' actions, which were unusual, and the
people interpreted that peculiarity as a reflection on theirs.

The Pythagoreans' kindred were indignant
that they associated with none, their parents
excepted; that they shared in common their possessions to the exclusion of their kindred,
whom they treated as strangers. These personal
motives turned the general opposition into active
hostility. Hippasus, Diodorus and Theages united
in insisting that the assembly and the magistracy should be opened to every citizen, and that
the rulers should be responsible to elected
representatives of the people. This was opposed
by the Pythagoreans Alcimachus, Dimachus, Meton
and Democedes, and opposed changes in the inherited constitution. They were however defeat-
ed, and were formally accused in a popular assembly by two orators, the aristocrat Cylon,
and the plebeian Ninon. These two planned their
speeches together, the first and longer one
being made by Cylon, while Ninon concluded by
pretending that he had penetrated the Pythagorean
mysteries, and that he had gathered and written
out such particulars as were calculated to crim-
inate the Pythagoreans, and he a scribe he gave
a book to read a book which was entitled the
Sacred Discourse.

from what Aristoxenus, in his treatise on the
Pythagoric Life says he heard from Dionysius
the tyrant of Sicily, when having been deposed
he taught language at Corinth. Here are the
words of Aristoxenus: "So far as they could
these men avoided lamentations and tears, and
the like; also adulation, entreaty, supplicati
and other emotions. Dionysius therefore, having
fallen from his tyranny and come to Corinth, t
us the detailed story about the Pythagoreans Ph
tias and Damon, who were sponsors for each othe
death. This is how it was: Certain intimates of
his had often mentioned the Pythagoreans, defam
and reviling them, calling them arrogant, and a
serting that their gravity, their pretended fid
ity, and stoicism would disappear on falling in
some calamity. Others contradicted this; and as
contention arose on the subject, it was decided
to settle the matter by an experiment. One man
accused Phintias, before Dionysius, of having
conspired with others against his life. Others
corroborated the charges, which looked probable,
though Phintias was astonished at the accusation.
When Dionysius had unequivocally said that he ha
verified the charges, and that Phintias must die
the latter replied that if Dionysius thought tha
this was necessary, he requested the delay of th
remainder of the day, to settle the affairs of
himself and Damon, as these two men lived togeth
and had all things in common; but as Phintias
was the elder, he mostly undertook the management of the household affairs. He therefore requested that Dionysius allow him to depart for
this purpose, and that he would appoint Damon a
his surety. Dionysius claimed surprise at such
a request, and asked him if any man existed who
would stand surety for the death of another. Phin
ias asserted that there was, and Damon was sent f
for; and on hearing what had happened, agreed to
become the sponsor, and that he would remain ther
till Phintias's return. Dionysius declared astonishment at these circumstances, and they who
had proposed the experimented derided Damon as
the one who would be caught, sneering at him as

the "vicarious stag;"when however sunset approached, Phintias came to die; at which all present were astonished and subdued. Dionysius, having embraced and kissed the men, requested that they would receive him as a third into their friendship. They however would by no means consent to anything of the kind, though he entreated them them to comply with his request. " These words are related by Aristoxenus, who received them from Dionysius himself.

It is also said, that the Pythagoreans endeavored to perform the offices of friendship to those of their sect, though they were unknown, and had never seen each other; on receiving a sure indication of participation in the same doctrines; so that, juging from such friendly offices, it may be believed, as is generally reported, that worthy men, even though they should dwell in the remotest parts of the earth, are mutually friends, and this before before they became known to, and salute each other.

The story runs that a certain Pythagorean, travelling through a long and solitary road on foot, came to an inn; and there from over-exertion, or other causes, fell into a long and severe disease, so as at length to want the necessaries of life. The inn-keeper, however, whether from pity or benevolence, supplied him with every thing requisite, sparing neither personal service, nor expense. Feeling the end near, the Pythagorean wrote a certain symbol on a tablet, and desired the inn-keeper, in event of his death, to hang the tablet near the road, and observe whether any traveller read the symbol. For that person, said, he, will repay you what you have spent on me, and will also thank you for your kindness. On the Pythagorean's death the innkeeper buried him, and attended to the obsequies, without any expectation of being repaid, no r of receiving any remuneration from anybody who might read the tablet. However, struck with the Pythagorean's request, he was induced to expose the writing in the public read. A long time thereafter a Pythagorean passed that way, and on understanding the

the existing regime; banishing along with them
all their families, on the two fold pretext that
impiety was unbearable, and that the children
should not be separated from their parents. They
then repudiated the debts, and redistributed
the lands.

Many years after, when Dinarchus and his associates had been slain in another battle, and whe°
Litagus, the chief leader of the sedition, were
dead, pity and repentance induced the citizens
to recall from exile what remained of the Pythagoreans. They therefore sent for messengers
from Achaia, who were to come to an agreement
with the exiles, and file their oaths (of loyal-
ty to the existing Crotonian regime ?) at Delphi.
The Pythagoreans who returned from exile were
about sixty in number, not to mention the aged,
among whom were some physicians and dieticians,
on original lines. When these Pythagoreans returned, they were welcomed by the crowds, who
silenced dissenters by announcing that the Ninon
regime was ended. Then the Thurians invaded the
country, and the Pythagoreans were sent to procure assistance;but they perished in battle, mutually defending each other. So thoroughly had
the city become Pythagoreanized that beside the
public praise, they performed a public sacrifice in the temple of the Muses which had originally been built at the instigation of Pythagoras.

That is all of the attack on the Pythagoreans.

CHAPTER XXXVI

THE PYTHAGOREAN SUCCESSION

Pythagoras's acknowledged successor was
Aristaeus, the son of the Crotonian Damophōn,
who was Pythagoras's contemporary, and lived
seven ages before Plato. Being exceedingly
skilful in Pythagoric dogmas, he succeeded the
school, educated Pythagoras's children, and mar‐
ried his wife Theano. Pythagoras was said to have
taught his school 39 years, and to have lived a
century. Aristaeus growing old, he relinquished t
the school to Pythagoras's son Mnesarchus. He was
followed by Bulagoras, in whose time Crotona was
plundered. After the war, Gartydas the Crotonian
who had been absent on a journey, returned,
and took up the school; but he so grieved about
his country's calamity that he died prematurely.
Pythagoreans who became very old were accustomed
to liberate themselves from the body, as a pris‐
on.

Later, being saved through certain strangers,
Aresas Lucanus undertook the school; and to him
came Diodorus Aspendius, who was received into
the school because of the small number of gen‐
uine Pythagoreans. C

Clinias and Philolaus were at Heraclea; Theo‐
Theorides and Eurytus at Metapontum, and at
Tarentum, Archytas. Epicharmus was alsosa said
to have been one of the foreign Hearers, but he
was not one of the school. However, having ar‐
rived at Syracuse, he refrained from public phil‐
osophizing, in consideration of the tyranny of
Hiero. But he wrote the Pythagorean views in
metre, and published the occult Pythagorean
dogmas in comedies.

It is probable that the majority of the Pythag‐
oreans were anonymous, and remain unknown. But
the following names are known and celebrated;

Of the Crotonians, Hippostratus, Dymas,
Aegon, Aemon, Sillus, Cleosthenes, Agelas, Epis‐
ylus; Phyciadas, Ecphantus, Timaeus, Buthius,
Eratus, Itmaeus, Rhodippus, Bryas, Evandrus,
Myllias, Antimedon, Ageas, Leophron, Agylus,

Onatus, Hipposthenes, Cleophron, Alcmaeon, Democles, Milon, Menon.

At Metapontum resided Brontinus, Parmiscus, Orestadas, Leon, Damarmenus, Aeneas, Chilas, Melisias, Aristeas, Laphion, Evandrus, Agesidamus, Xenocades, Euryphemus, Aristomenes, Agesarchus, Alceas, Xenophantes, Thraseus, Arytus, Epiphron, Eiriscus, Megistias, Leocydes, Thrasymedes, Euphemus, Procles, Antimenes, Lacritus, Damotages, Pyrrho, Rhexibius, Alopecus, Astylus, Dacidas, Aliochus, Lacrabes and Glycinus.

Of the Agrigentines was Empedocles.

Of the Eleatae, was Parmenides.

Of the Tarentines were Philolaus, Eurytus, Archytas, Theodorus, Aristippus, Lycon, Hestiyaeus, Polemarchus, Asteas, Clinias, Cleon, Eurymedon, Arceas, Clinagoras, Archippus, Zopyrus, Euthynus, Dicearchus, Philonidas, Phrontidas, Lysis, Lysibius, Dinocrates, Echecrates, Paction, Acusilidas, Icmus, Pisicrates, and Clearatus.

Of the Leontines were Phrynichus, Smichias, Aristoclidas, Clinias, Abroteles, Pisyrrhydus, Bryas, Evandrus, Archemachus, Mimnomachus, Achmonidas, Dicas Dicas and Carophantidas.

Of the Sybarites were Metopus, Hippasus, Proxenus, Evanor, Deanax, Menestor, Diocles, Empedus, Timasius, Polemaeus, Evaeus, and Tyrsenus.

Of the Carthaginians was Miltiades, Anthen, Odius and Leocritus.

Of the Parians, Aeetius, Phaenecles, Dexitheus, Alchimachus, Dinarchus, Meton, Timaeus, Timesianax, Amaerus, and Thymaridas.

Of the Locrians, Gyptius, Xenon, Philodamus, Evetes, Adicus, Sthenonidas, Sosistratus, Euthynus, Zaleucus, Timares.

Of the Posidonians, Athamas, Simus, Proxenus, Cranous, Myes, Bathylaus, Phaedon.

Of the Lucani, Ocellus, and his brother Occillus, Oresandrus, Cerambus, Dardaneus, and Malion.

Of the Aegeans, Hipponedon, Timosthenes, Euelthon, Thrasydamus, Crite, and Polyctor.

Of the Hyperboreans, Abaris.

Of the Lacones, Autocharidas, Cleanor, Eurycrates.

Of the Rheginenses, Aristides, Demosthenes, Aristocrates, Phytius, Helicaon, Mnesibulus, Hipparchides, Athosion, Euthycles, Opsimus.

Of the Selinuntians, Calais.

Of the Syracusans, Leptines, Phintias, and Damon.

Of the Samians, Melissus, Lacon, Archippus, Glorippus, Heloris, Hippon.

Of the Caulonienses, Callibrotus, Dicon, , Nastas, Drymen and Xentas.

Of the Phliasians, Diocles, Echecrates, Phanton, and Polymnastus.

Of the Sicyonians, Poliades, Demon, Sostratius, and Sosthenes.

Of the Cyrenians, Prorus, Melanippus, Aristangelus, and Theodorus.

Of the Cyziceni, Pythodorus, Hipposthenes, Butherus and Xenophilus.

Of the Catanaei, Charondas and Lysiades.

Of the Corinthians, Chrysippus.

Of the Tyrrhenians, Nausitheus .

Of the Athenians, Neocritus.

Of the Pontians, Lyramnus.

In all, two hundred and eighteen.

The most illustrious Pythagorean women are Timycha, the wife of Myllias the Crotonian; Phyltis, the daughter of Theophrius the Crotonian. Byndacis, the sister of Ocellus and Occillus, Lucanians. Chilonis, the daughter of Chilon the Lacedemonian. Cratesiclea the Lacedemonian, the wife of the Lacedemonian Cleanor. Theano, the wife of Brontinus of Metapontum. Mya, the wife of Milon the Crotonian. Lasthenia the Arcadian. Abrotelia, the daughter of Abroteles the Tarentine. Echecratia the Phliasian. Tyrsenis the Sybarite; Pisirrhonde, the Tarentine. Nisleadusa, the Lacedemonian. Bryo, the Argive. Babelyma the Argive, and Cleaechma, the sister of Autocharidas the Lacedemonian. In all, seventeen.

LIFE OF PYTHAGORAS

by PORPHYRY

1. Many think that Pythagoras was the son of
Mnesarchus, but they differ as to the latter's
race; some thinking him a Samian, while Neanthes,
in the fifth book of his Fables states he was a
Syrian, from the city of Tyre. As a famine had
arisen in Samos, Mnesarchus went thither to trade,
and was naturalized. there. There also was born
his son Pythagoras, who early manifested studiousness, but was later taken to Tyre, and there entrusted to the Chaldeans, whose doctrines he im-
bibed. Thence he returned to Ionia, where he first
studied under the Syrian Pherecydes, then also
under Hermodamas the Creophylian who at that time
was an old man, residing in Samos.

 . 2. Neanthes says that others hold that his father was a Tyrrhenian, of those who inhabit Lemnos, and that while on a trading trip to Samos was
there naturalized. On sailing to Italy, Mnesarchus
took the youth Pythagoras with him. Just at this
time this country was greatly flourishing. Neanthes
adds that Pythagoras had two older brothers, Eunos-
tus and Tyrrhenus. But Apollonius, in his book about Pythagoras, affirme that his mother was Pythais,
a descendant of Ancaeus, the founder of Samos.
Apollonius adds that he was siad to be the offspring of Apollo and Pythais, on the authority of
Mnesarchus; and a Samian poet sings:
   "Pythais, of all Samians the most fair,
    Jove-loved Pythagoras to Phoebus bare!"
This poet says that Pythagoras studied not only
under Pherecydes and Hermodamas, but also under
Anaximander.

3. The Samian Duris, in the second book of his
"Hours," writes that his son was named Arimnestus,
that he was the teacher of Democritus, and that
on returning from banishment, he suspended a bras-
######### in the temple of Hera, a tablet two
feet square, bearing this inscription:

"Me, Arimnestus, who much learning traced,
  Pythagoras's beloved son here placed."
This tablet was removed by Simus, a musician, who
claimed the canon graven thereon, and published it
as his own. Seven arts were engraved, but when
Simus took away one, the others were destroyed.

4. It is said that by Theano, a Cretan, the
daughter of Pythonax, he had a son, Telauges,
and a daughter, Myia; to whom some add Arignota,
whose Pythagoric writings are still extant. Timaeus
relates that Pythagoras's daughter, while a maid-
en, took precedence among the maidens in Crotona,
and when a wife, among married women. The Crotonians
made her house a temple of Demeter, and the neighboring street they called a museum.

5. Lycus, in the fourth book of his Histories,
noting different opinions about his country, says,
"Unless you happen to know the country and the
city of which Pythagoras was a citizen, will remain a mere matter of conjecture. Some say he was
a Samian, others, a Phliasian, others a Metapontine."

6. As to his knowledge, it is said that he
learned the mathematical sciences from the Egyptians, Chaldeans and Phoenicians; for of old the Eg
Egyptians excelled in geometry, the Phoenicians
in numbers and proportions, and the Chaldeans of
astronomical theorems, divine rites, and worship
of the Gods; other secrets concerning the course
of life he received and learned from the Magi.

7. These accomplishments are the more generally known, but the rest are less celebrated. Moreover Eudoxus, in the second book of his Description of the Earth, writes that Pythagoras used the
greatest purity, and was shocked at all bloodshed-
ding and killing; that he not only abstained from
animal food, but never in any way approached
butchers or hunters. Antiphon, in his book on
Illustrious Virtueus Men praises his perseverance
while he was in Egypt, saying, "Pythagoras desiring to become acquainted with the institutions of

the Egyptian priests, and diligently endeavoring
to participate therein, requested the Tyrant Polycrates to write to Amasis, the King of Egypt, his
friend and former host, to procure him initiation.
Coming to Amasis, he was given letters to the prie
priests; of Heliopolis, who sent him on to those ß
of Memphis, on the pretense that they were the mor
more ancient. On the same pretense, he was sent on
from Memphis to Diospolis.

8. From fear of the King, the latter priests
dared not make excuses; buy thinking that he would
desist from his purpose as result of great difficulties, enjoined on him very hard precepts, entirely different from the institutions of the Greeks,
These he performed so readily that he won their
admiration, and they permitted him to sacrifice
to the Gods, and to acquaint himself with all
their sciences, a favor theretofore never granted
to a foreigner.

9. Returning to Ionia, he opened ~~########~~ in
his own country, a school, which is even now called Pythagoras's Semicircles, in which the Samians
meet to deliberate about matters of common interest. Outside the city he made a cave adapted to
the study of his philosophy, in which he abode
day and night, discoursing with a few of his associates. He was now fourty years old, says Aristoxenus. Seeing that Polycrates's government was becoming so violent that soon a free man would become a victim of his tyranny, he journeyed towards
Italy.

10. Diogenes, in his treatise about the Incred-
ible Things Beyond Thule, has treated Pythagoras's
affairs so carefully, that I think his account
should not be omitted. He says that the Tyrrhenian
Mnesarchus was of the race of the inhabitants of
Lemnos, Imbros and Scyros; and that he departed
thence to visit many cities and various lands.
During his journeys he found an infant lying under a large, tall poplar tree. On approaching, he
observed it lay on its back, looking steadily
without winking at the sun. In its mouth was a

little slender reed, like a pipe; through which
the child was being nourished by the dew-drops
that distilled from the tree. This great wonder
prevailed upon him to take the child, believing
it to be of a divine origin. The child was fos-
tered by a native of that country, named Androcles, who later on adopted him, and entrusted to
him the management of his affairs. On becoming
wealthy, Mnesarchus educated the boy, naming him
Astraeus, and rearing him with his own three
sons, Eunestus, Tyrrhenus, and Pythagoras; which
boy, as I have said, Androcles adopted.

11. He sent the boy to a lute-player, a wrest
ler and a painter. Later he sent him to Anaximander
der at Miletus, to learn geometry and astronomy.
Then Pythagoras visited the Egyptians, the Arabi-
ans, the Chaldeans and the Hebrews, from whom he
acquired expertry in the interpretation of dreams,
and he was the first to use frankincense in the
worship of divinities.

12. In Egypt he lived with the priests, and
learned the language and wisdom of the Egyptians,
and their three kinds of letters, the epistolic,
the hieroglyphic, and symbolic, whereof one imitates
ates the common way of speaking, while the ethers
express the sense by allegory and parable. In Ara-
bia he conferred with the King. In Babylon he
associated with the other Chaldeans, specially
attaching himself to Zabratus, by whom he was
purified from the pollutions of his past life,
and taught the things from which a virtuous man
ought to be free. Likewise he heard lectures about Nature, and yhe principles of wholes. It was
from his stay among these foreigners that Pythagoras acquired the greater part of his wisdom.

13. Astraeus was by Mnesarchus entrusted to
Pythagoras, who received him, and after studying
his physionomy and the motions of his body, instructed him. First he accurately investigated
the science about the nature of man, discerning th
the disposition of every one he met. None was allowed to become his friend or associate without
being examined in facial expression and disposition.

14. Pythagoras had another youthful disciple, from Thrace. Zamolxis was he named bevause he was born wrapped in a bear's skin, in Thracian called zalmus. Pythagoras loved him, and instructed him in sublime specualtions concerning sacred rites, and the natuuepof the Gods. Some say this youth was named Thales, and that the barbarians worshipped him as Hercules.

15. Dionysiphanes says that he was a servant of Pythagoras, who fell into the hands of thieves and by them was branded. Then when Pythagoras was persecuted and banished, (he followed him) binding up his forehead on account of the scars. Others say that the name Zamolxis signifies a stranger or foreigner.

Pherecydes, in Delos, fell sick; and Pythagoras attended him until he died, and performed his funeral rites. Pythagoras then, longing to be with Hermodamas the Creophylian, returned to Samos. After enjoying his society, Pythagoras trained the Samian athlete Eurymenes, who though he was of small stature, conquered at Olympia, through his surpassing knowledge of Pythagoras's wisdom. While according to ancient custom the ither athletes fed on cheese and figs, Eurymenes, by the advice of Pythagoras, fed daily on flesh, which endued his body with great strength. Pythag-#men##ghadually imbued him with his wisdom, exhorting him to go into the struggle, not for the sake of victory, but the exercise; that he should gain by the training, avoiding the envy resulting from victory. For the victors, are not always pure, though decked with leafy crowns.

16. Later, when the Samians were oppressed with the tyranny of Popycrates, Pythagoras saw that life in such a state was unsuitable for a philosopher, and so planned to travel to Italy. At Delphi he inscribed an elegy on the tomb of Apollo, declaring that Apollo was the son of Silenus, but was slain by Pytho, and placed in the place called Triops, so named from the local mourning for Apollo by the three daughters of Triopas.

17. Going to Crete, Pythagoras besought initiation from the priests of Morgos, one of the
Idaean Dactyli, by whom he was purified with the
meteoritic thunder-stone. In the morning he lay
stretched upon his face by the sea-side; at night,
he lay beside a river, crowned with a black lamb's
wooolen wreath. Descending into the Idaean cave,.
wrapped in black wool, he stayed there twenty-
seven days, according to custom; he sacrificed to
zeus, and saw the throne which there is yearly made
for him. On Zeus's tomb, Pythagoras inscribed an
epigram, "Pythagoras to Zeus," which begins:
"Zeus deceased here lies, whom men call Jove."

18. When he reached Italy, he stopped at
Crotona. His presence was that of a free man, tall,
graceful in speech and gesture, and in all things
else. Dicaearchus relates that the arrival of
this great ############traveller, endowed with
all the advantages of nature, and prosperously
guided by fortune, produced on the Crotonians so
great an impression, that he won the esteem of the
the older magistrates, by his many and excellent
discourses. They ordered him to exhort the young
men, and then to the boys who flocked out of the
school to hear him; and lastly to the women, who
came together on purpose.

19. Through this he achieved great reputation,
and he drew great audiences from the city, not
only of men, but also of women, among whom was
a specially illustrious person named Theano. He
also drew audiences from among the neighboring
barbarians, among whom were magnates and kings.
What he told his audiences cannot be said with
certainty, for he enjoined silence upon his hearers. But the following is a matter of general information. He taught that the soul was immortal,
and that after death it transmigrated into other
animated bodies. After certain specified periods,
the same events occur again; that nothing was
entirely new; that all animated beings were kin,
and should be considered as belonging to one
great family. Pythagoras was the first one to
introduce these teachings into Greece.

20. His speech was so persuasive that, according to Nicomachus, in obe address made on first landing in Italy he made more than two thoudand adherents. Out of desire to live with him, these buiitwerhangmiandditorium, to which both women and built a large auditorium, to which both women and boys were admitted. (Foreign visitors were so many that) they built whole cities, settling that whole region of Italy now known as Magna Grecia. His ordinances and laws were by them received as divine precepts, and without them would do nothing. Indeed they ranked him among the divinities. They held all property in common. holing all their property in common. They ranked him among the divinities, and whenever they communicated to each other some choice bit of his philosophy, from which physical truths could always be deduced, they would swear by the Tetractys, adjuring Pythagoras as a divine witness, in the words,

I call to witness him who to our souls expressed
The Tetractys, eternal Nature's fountain-spring."

21. During his travels in Italy and Sicily he foundnvarious cities subjected one to another, both of long standing, and recently. By his disciples, some of whom were found in every city, he infused into them an aspiration for liberty; thus restoring to freedom Crotona, Sybaris, Catana, Rhegium, Himera, Agrigentum, Tauromenium, and others, on whom he imposed laws through Charondas the Catanean, and Zaleucus the Locrian, which resulted in a long era of good government, emulated by all their neighbors. Simichus the tyrant of the Centorupini, on hearing Pythagoras discourse, abdicated his rule, and divided his property between his sister and the citizens.

22. According to Aristoxenus, some Lucanians, Messapians, Picentinians and Romans came to him. He rooted out all dissensions, not only among his de discipkes and their successors, for many ages, but among all the cities of Italy and Sicily, both internally and edternally. He was continuously harpig on this maxim, "We ought, to the best of our ability avoid, and even with fire and sword extirpate frm

the body, sickness; fromnthe soul, ignorance; from
the belly, luxury; from a city, sedition; from a
family, discord; and from all things, excess."

23. Ifwe may credit what ancient and trustworthy writers have related of him, he exerted an
influence even over irrational animals. The Daunian bear, who had commited extensive depredations
in the neighborhood, he seized; and after having
patted her for awhile, and given her barley and
fruits, he made her swear never again to touch
a living creature, and then released her. She immediately his herself in the woods and the hills,
and from that time on never attacked any irrational animal.

24. At Tarentum, in a pasture, seeing an ox
cropping beans, he went to the herdsman, and adv
vised him to tell the ox to abstain from beans.
The countryman mocked him, proclaiming his ignorance of the ox-language. So Pythagoras himself
went and whispered in the ox's ear, Not only did
the bovine at once desist from his diet of beans,
but would never touch any thenceforward, though
he survived many years near Hera's temple at Tarentum, until very old; being called the sacred
ox, and eating any food given him.

25. While at the Olympic games, he was disec___
coursing with his friends about auguries, omens,
and divine signs, and how men of true piety do
receive messages from the Gods. Flying over his
head was an eagle, who stopped, and came down to
Pythagoras. After stroking her awhile, he released
her.
Meeting with some fishermen who were drawing
in their nets heavily laden with fishes from the
deep, he predicted the exact number of fish they
had caught. The fishermen said that if his estimate was accurate they would di whatever he commanded. They counted them accurately, and found the
number correct. He then bad them return the fish
alive into the sea; and, what is more wonderful,
not one of them died, although they had been out
of the water a considerable time. He paid them
and left.

26. Many of his associates he reminded of the lives lived by their souls before it was bound to the body, and by irrefutable arguments demonstrated that he had been Euphorbus, the son of Panthus. He specially praised the following verses about himself, and sang them to the lyre most elegantly:

"The shining circlets of his golden hair,
Which even the Graces might be proud to wear,
Instarred with gems and gold, bestrew the shore,
With dust dishonored, and deformed with gore.
As the young olive, in some sylvan scene,
Crowned by fresh fountains with celestial green,
Lifts the gay head in snowy flowerets fair,
And plays and dances to the gentle air,
When lo, a whirlwind from high heaven invades,
The tender plant, and withers all its shades;
It lies uprooted from its genial head,
A lovely ruin, now defaced and dead.
Thus young, thus beautiful, Euphorbus lay,
While the fierce Spartan tore his arms away."
                        (Pope, Homer's Iliad, Book 17).

27. The stories about the shield of this Phrygian Euphorbus being at Mycenae dedicated to Argive Hera, along with other Trojan spoils, shall here be omitted as being of too popular a nature.

It is said that the river Caicasus, while he, with many of his associayes was passing over it, spoke to him very cleary, "Hail, Pythagoras!". Al Almost unanimous is the report that on one and the same day he was present at Metapontum in Italy, and at Tauromenium in Sicily, in each place conversing with his friends, though the places are separated by many miles, both at sea and land, demanding many days' journeys.

28. It is well known that he showed his golden thigh to Abaris the Hyperborean, to confirm him in the opinion that he was the Hyperborean Apollo, whose priest Abaris was.

A ship was coming into the harbor, and his friends expressed the wish to own the goods it contained. "Then," said Pythagoras, "you would own a corpse!" On the ship's arrival, this was found to be the true state of affirs.

Of Pythagoras many other more wonderful and
divine things are persistently and unanimously
related, so that we have no hesitation in saying
never was more attributed to any man, nor was any
more eminent.

29. Verified predictions of earthquakes are
handed down, also, that he immediately chased away
a pestilence, suppressed violent winds and hail,
calmed storms both on rivers and on seas, for
the comfort and safe passage of his friends. As
their poems attest, the like was often performed
by Empedocles, Epimenides and Abaris, who had
learned the art of doing these things from him.
Empedocles, indeed, was surnamed Alexanemos, as
the chaser of winds; Epimenides, Cathartes, the
Lustrator. Abaris was called Aethrobates, the
walker in air; for he was carried in the air on
an arrow of the Hyperborean Apoollo, over rivers,
seas and inaccessible places. It is believed that
this was the method employed by Pythagoras when
on the same day he discoursed with his friends
at Metapontum and Tauromenium.

30, He soothed the passions of the soul and
body by rhythms, songs, and incantations. These
he adapted and applied to his friends. He himself
could hear the harmony of the Universe, and understood the universal music of the spheres, and of
the stars which move in concert with them, and
which we cannot hear because of the limitations
of our weak nature. This is testified to by these
characteristic verses of Empedocles:
"Amongst these was one in things sublimest skilled,
His mind with all the wealth of learning filled."
Whatever sages did invent, he sought;
And whilst his thoughts were on this work intent,
All things existent, easily he viewed,
Through ten or twenty ages making search."

31. Indicating by sublimest things, and, he
surveyed all existent things, and the wealth of
the mind, and the like, Pythagoras's constitution of body, mind, seeing, hearing and understanding, which was exquisite, and surpassingly accurate.

Pythagoras affirmed that the Nine Muses were
constituted by the sounds made by the seven planets, the sphere of the fixed stars, and that which
is opposed to our earth, called "anti-earth." He
called Mnemosyne, or Memory, the composition,
symphony and connexion of them all, which is eter
nal and unbegotten as being composed of all of them.

32. Diogenes, setting forth his daily routine
of living, relates that he advised all men to
avoid ambition and vain-glory, which chiefly excite envy, and to shun the presences of crowds.
He himself held morning conferences at his residence, composing his soul with the music of the
lute, and singing certain old paeans of Thales.
He also sang verses of Homer and Hesiod, which
seemed to soothe the mind. He danced certain
dances which he conceived conferred on the body
agility and health. Walks he took not promiscuously, but only in company of one or two companions,
in temples or sacred groves, selecting the quietes
and pleasantest places.

33. His friends he loved exceedingly, being
the first to declare that the goods of friends
are common, and that a friend was another self.
While they were in good health he always conversed with thm; if they were sick, he nursed them;
if they were afflicted in mind, he solaced them,
some by incantations and magic charms, others by
music. He had prepared songs for the diseases of
the body, by the singing of which he cured the
sick. He had also some that caused oblivion of
sorrow, mitigation of anger, and destruction of
lust.

34. As to food, his breakfast was chiefly of
honey; at dinner he used bread made of millet,
barley or herbs, raw and boiled. Only rarely did
he eat the flesh of victims; nor did he take this
from every part of the anatomy. When he intended
to sojourn in the sanctuaries of the divinities,
he would eat no more than was necessary to still
hunger and thirst. To quiet hunger, he made a
mixture of poppy seed and sesame, the skin of a
sea-onion, well washed, till entirely drained of

THE OUTWARD JUICE; of the flowers of the daffodil,
and the leaves of mallows, of paste of barley and
pea; taking an equal weight of which, and chopping
it small, with Hymettian honey he made it into s
mass. Against thirst he took the seed of cucumbers,
and the best dried raisins, extracting the seeds,
and the flower of cariander, and the seeds of mallows, purselain, scraped cheese, meal and cream;
these he made up with wild honey.

b   35. He claimed that this diet had, by Demeter,
been taught to Hercules, when he was sent into the
Lybian deserts. This preserved his body in an
unchanging condition; not at one time well, and at
another time sick, nor at one time fat, and at
another lean. Pythagoras's countenance showed
the same constancy was in his soul also. For he w
was neither more elated by pleasure, nor dejected
by grief; and no one ever saw him either rejoicing or mourning.

36. When Pythagoras sacrificed to the Gods,
he did not use offensive profusion, but offered
no more than barley bread, cakes and myrrh; least
of all animals, unless perhaps cocks and pigs.
When he discovered the proposition that the square
on the hypothenuse of a right angled triangle was
equal to the squares on the sides containing the
right angle,he is said to have sacrificed an ox,
although the more accurate say that this ox was
made of flour.

37. His utterances were of two kinds, plain
or symbolical. His teaching was twofold: of his
disciples some were called Students, and others
Hearers. The Students learned the fuller and
more exactly elaborate reasons of science, while
the Hearers heard only the chief heads of learning, without more detailed explanations.

38. He ordained that his disciples should
speak well and think reverently of the Gods,
geniuses and heroes, and likewise of parents and
benefactors; that they should obey the laws; that
should not relegate the worship of the Gods to

a secondary position, performing it eagerly, even at home; that to the celestial divinities they should sacrifice uncommon offerings, and ordinary ones to the inferior deities. (The world he divided into) opposite powers; the "one" was a better monad, light, right, equal, stable and straight; while the "other" was an inferior duad, darkness, left, unequal, unstable and movable.

39. Moreover, he enjoined the following. A cultivated and fruit-bearing plant, harmless to man and beast, should be neither injured nor destroyed. A deposit of money or of teachings should be faithfully preserved by the trustee.

There are three kinds of things that deserve to be pursued and acquired: honorable and virtuous things, those that conduce to the use of life, and those that bring pleasures of the blameless, solid and grave kind, of course not the vulgar intoxicating kinds. Of pleasures there were two kinds: one that indulges the bellies and lusts by a profusion of wealth, which he compared to the murderous songs of the Sirens; the other kind consists of things honest, just, and necessary to life, which are just as sweet as the first, without being followed by repentance; and these pleasures he compared to the harmony of the Muses.

40. He advised special regard to two times: that when we go to sleep, and that when we awake. At each of these we should consider our past actions, and those that are to come. We ought to require of ourselves an account of our past deeds, while of the future we should have a providential care. Therefor he advised everybody to repeat to himself the following verses before he fell asleep:
"Nor suffer sleep to close thine eyes
 Till thrice thy acts that day thou hast run o'er;
 How slipt?What deeds? What duty left undone?"
On rising:
 "As soon as ere thou wakest, in order lay
 The actions to be done that following day."

41. Such things taught he, though advising
above all things to speak the truth, for this
alone deifies men. For as he had learned from
the Magi, who call God Oromasdes, God's body is
light, and his soul is truth. He taught much else,
which he claimed to have learned from Aristoclea at
at Delphi. Certain things he declared mystically,
symbolically, most of which were collected by Aristotle, as when he called the sea a tear of Saturn; the two bear (constellations) the hand of
Rhea; the Pleiades, the lyre of the Muses;the
planets, the dogs of Persephone; and he called
the sound caused by striking on brass the voice
of a genius enclosed in the brass.

42.   He had also another kind of symbols, such
such as, Pass not over a balance; that is, Shun
avarice. Poke not the fire with a swird, that is, we
we ought not to excite a man full of fire and anger with sharp language. Pluck not a crown meant
not toviolate the laws, which are the crowns of
cities. Eat not the heart, signified not to afflict
flict ourselves with sorrows. Do not sit upon a
peck-measure, meant, do not live ignobly. On starting a journey, do not turn beck, meant that this
life should not be regretted, when near the bourne of
of death. Do not walk in the public way meant to
avoid the opinions of the multitude, adopting
those of the learned and the few; Receive not
swallows into your house, meant not to admit
under the same roof garrulous and intemperate
men. Help a man to take up a vurden, but not to
lay it down, meant to encourage no one to be indolen
dolent, but to apply oneself to labor and virtue.

Do not carry the images of the Gods in rings, signified that one should not at once to the vulgar
reveal one's opinions about the Gods, or discourse
about them. Offer libations to the Gods, just to
the ears of the cup, meant that we ought to worship
and celebrate the Gods with music, for that penetrates through the ears. Do not eat those things that
that are unlawful, sexual, or increase, beginningnor
er end, nor the first basis of all things.

43. He taught abstention from the loins,
testicle, pudenda, marrow, feet and heads of
victims. The loins he called <u>basis,</u> because on
them as foundations living beings are settled.
Testicles and pudenda he called <u>generation,</u>
for no one is engendered without the help of these.
Marrow he called <u>increase,</u> as it is the cause of
growth in living beings. The <u>beginning</u> was the
feet, and the head <u>the end;</u>which have the most
power in the government of the body. He likewise
advised abstention from beans, as from human flesh.

44. Beans were interdicted, it is said, because
the particular plants grow and individualize only
after (the earth) which is the principle and origin of things, is mixed together, so that many
things underground are confused, and coalesce;
after which everything rots together. Then living
creatures were produced together with plants, so
that both men and beans arose out of putrefaction;
whereof he alleged many manifest arguments. For
if any one should chew a bean , and having ground
it to a pulp with his teeth, and should expose
that pulp to the warm sun, for a short while, and
then return to it, he will perceive the scent of
human blood. Moreover, if at the time when beans
bloom, one should take a little of the flower, w
which then is black, and should put it into an
earthen vessel, and cover it closely, and bury i
in the ground for ninety days, and at the end th
thereof take it up, and uncover it, instead of t
bean he will find either the head of an infant,
or the pudenda of a woman.

45. He also wished men to abstain from other
things, such as a swine's paunch, a mullet, and
a sea-fish called a"nettle," and from nearly all
other marine animals. He referred his origin to
those of past ages, affirming that he was first
Euphorbus, then Aethalides, then Hermotimus, then
Pyrrhus, and last, Pythagoras. He showed to his
disciples that the soul is immortal, and to those
who were rightly purified he brought back the mem
ory of the acts of their former lives.

46. He cultivated philosophy, the scope of
which is to free the mind implanted within us
from the impediments and fetters within which it
is confined; without whose freedom none can learn
anything sound or true, or perceive the unsoundness
in the operation of sense. Pythagoras thought that
mind alone sees and hears, while all the rest are bli
sblind and deaf. The purified mind should be applied
to the discovery of beneficial things, which can
be effected by certain artificial ways, which by
degrees induce it to#the contemplation of etern-   —
al and incorporeal things, which never vary. This
orderliness of perception should begin from consideration of the most minute things, lest by any
changethe mind should be jarred and withdraw itself, through the ######### failure of contin-   —
uousness in its subject-matter.
    47.

47. That is the reason he made so much use of
the mathematical disciplines and speculations,
which are intermediate between the physical and
the incorporeal realm, for the reason that like
bodies they have a three-fold dimensions, and yet
share the impassibility of incorporeals; as degrees of preparation to the contemplation of the
really existent things; by an artificial reason
diverting the eyes of the mind from corporeal
things, whose manner and state never remain in
the same condition, to a desire for true (spiritual) food; By means of these mathematical
sciences therefore, Pythagoras rendered men truly
happy, by this artistic introduction of truly
existent things.

48. Among others, Moderatus of Gades, who
learnedly treated of the qualities of numbers in
eleven books, states that the Pythagoreans specialized in the study of numbersto explain their ######
teachings symbolically, as do geometricians, inasmuch as the primary forms and principles are hard
to understand and express,otherwise, in plain discourse. A similar case is the representation of
sounds by letters, which are known by marks, which
are called the first elements of learning; later,

they inform us these are not the true elements,
which they only signify.

49. As the geometricians cannot express incorporeal forms in words, and have recourse to
the descriptions of figures, as that & is a triangle, and yet do not mean that the actually seen
lines are the triangle, but only what they represent, the knowledge in the mind, so the Pythagoreans used the same objective method in respect
to first reasons and forms. As these incorporeal
forms and first principles could not be expressed in words, they had recourse to demonstration
by numbers. Number one denoted to them the reason of Unity, Identity, Equality, the purpose of
friendship, sympathy, and conservation of the Universe, which results from persistence in Sameness.
For unity in the details harmonizes all the parts
of a whole, as by the participation of the First
Cause.

50. Number two, or Duad, signified the twofold reason of diversity and inequality, of everything that is divisible, or mutable, existing at
one time in one way, and at another time in another way. After all these methods were not confined to the Pythagoreans, being used by other
philosophers to denote unitive powers, which contain all things in the universe, among which are
certain reasons of equality, dissimilitude and
diversity. These reasons are what they meant by
the terms Monad and Duad, or by the words uniform,
biform, or diversiform.

50. The same reasons apply to their use of
other numbers, which were ranked according to
certain powers. Things that had a beginning, middle and end they denoted by the number Three, saying that anything that has a middle is triform,
which was applied to every perfect thing. They said
that if anything was perfect it would make use of
this principle, and be adorned according to it;
and as they had no other name for it, they invente
the form, Triad; and whenever they tried to bring
us to the knowledge of what is perfect they led

US TO That by the form of this Triad. So also with th with the other numbers, which were ranked according to the same reasons.

52. All other things were comptehended under a single form and power, which they called Decad, explaining it by a pun, as dechad, meaning comprehension. That is why they called Ten a perfect number, the most perfect of all, as comprehending all difference of numbers, reasons, species and proportions. For if the nature of the universe be defined according to the reasons and proportions of numbers, and if that which is produced, inereascreased and perfected, proceed according to the re reason of numbers; and since the Decad comprehends every reason of numbers, every proportion, and every species, — why should Nature herself not be denoted by the most perfect number, Ten? Such was the use of numbers among the Pythagoreans.

53. This primary philosophy of the Pythagoreans finally died out first, because it was enigmatical, and then because their commentaries were written in Doric, which dialect itself is some somewhat obscure, so that Doric teaffhings were not fully understood, and they became misapprehended, and finally spurious, and later, they who published them no longer were Pythagoreans. The Pythagoreans affirm that Plato, Aristotle, Speusi Speusippus, Aristoxenus and Xenocrates appropriated ted the best of them, making but minor changes. (To distract attention from this their theft), they later collected and deluvered as characterist eristic Pythagorean doctrines whatever therein was most trivial, and vulger, and whatever had been invented by envious and calumnious persons, to cast contempt on Pythagoreanism.

54. Pythagoras and his associates were long held in such admitation in Italy, that many cities invited them to undertake their administration. At last, however, they incurred envy, and a conspiracy was formed against them as follows. Cylo, a Crotonian, who in race, nobility and wealth

was th most preeminent, was of a severe, violent
and tyrannical disposition, and did not scruple
to use the multitude of his followers to compass
his ends. As he esteemed himself worthy of whatever was best, he considered it his right to be
admitted to Pythagorean fellowship. He therefore
went to Pythagoras, extolled himself, and desired
his conversation. Pythagoras, however, who was
accustomed to read in human bodies'# nature and
manners the disposition of the man, bade him depart, and go about his business. Cylo, being of
a rough and violent disposition, took it as a great
great affront, and became furious.

55. He therefore assembled his friends, began
to accuse Pythagoras, and conspired against him
and his disciples. Pythagoras then went to Delos,
to visit the Syrian Pherecydes, formerly his
teacher, who was dangerously sick, to nurse him.
Pythagoras's friends then gathered together in
the house of Milo the wrestler; and were all
stoned and burned when Cylo's followers set the
house on fire. Only two escaped, Archippus and.
Lysis, according to the account of Neanthes.
Lysis took refuge in Greece, with Epaminondas,
whose teacher he had formerly been.

56. But Dicaearchus and other more accurate
historians relate that Pythagoras himself
was present when this conspiracy bore fruit,
for Pherecydes had died before he left Samos.
Of his friends, forty who were gathered together in a house were attacked and slain; while
others were gradually slain as they came to the
city. As his friends were taken, Pythagoras himself first escaped to the Caulonian haven, and
thence visited the Locrians. Hearing of his coming, the Locrians sent some old men to their
frontiers to intercept him. They said, " Pythagoras, you are wise and of great worth; but as our
laws contain nothing reprehensible, we will preserve them intact. Go to some other place, and we will
furnish you with any needed necessaries of travel."

Pythagoras turned back, and sailed to Tarentum,
where, receiving the same treatment as at Crotona, he went to Metapontum. Everywhere arose great
mobs against him, of which even now the inhabitants
make mention, calling them the Pythagorean riots,
as his followers were called Pythagoreans.

57. Pythagoras fled to the temple of the
Muses, in Metapontum, There he abide forty days,
and starving, died. Others however state that
his death was due to grief at less of all his
friends who, when the house in which they were
gathered was burned, in order to make a way for
their master, they threw themselves into the
flames, to make a bridge of safety for him, whereby
by indeed he escaped. When died the Pythagoreans,
with them also died their knowledge, whixh till then
then they had kept secret, excpet for a few obscure thingsnwhich were commonly repeated by those
who did not understand them. Pythagoras himself
left no book; but some little sparks of his philosophy, obscure and difficult, were preserved by
the few who were preserved by being scattered, as
vere Lysis and Archippus.

58. The Pythagoreans now avoided human society,
being lonely, saddened and dispersed. Fearing nevertheless that among men the name of philosophy would
be entirely extinguished, and that therefore the
Gods would be angry with them, they made abstracts
and commentaries. Each man made his own collection
of written authorities and his own memories, leaving them wherever he happened to die, charging
their wives, sons and daughters to presrve them
within their families. This mandate of transmission
within each family was obeyed for a long time.

58. Nichomacus says that this was the reason
why the Pythagoreans studiously avoided friendship
with strangers, preserving a constant friendship
among each other.
Aristoxenus, in his book on the Life of Pythagoras, says he heard many things from Dionysius,
the tyrant of Sicily, who, after his abdication,
taught letters at Corinth. Among these were that

they abstained frem lamentations and grieving,
and tears; alse from adulation, entreaty, supplication and the like.

60. It is said that Dionysius at one time
wanted te test their mutual fidelity under impri-
sonment. He contrived this plan. Phintias was arrested, and taken before the tyrant, and charged
with plotting against the tyrant, convicted, and
condemned te death. Phintias, accepting the situation, asked to be given the rest of the day to
arrange his own affairs, and those of Damon, his
friend and associate, who now would have to assume the management. He therefore asked for a
temperaty release, leaving Damcn as security for
his appearance. Dionysius granted the request,
and they sent for Samon, who agreed to remain
until Phintias should return.

61. The novelty of this deed astonished Dien-
ysius; but those who had first suggested the
experiment, scoffed at Damon, saying he was in
danger of losing his life. But to the general
surprise, near sunset Phintias came to die. Dien-
ysius then expressed his admiration, embraced the
them beth, and asked to be received as a third
in their friendship. Though he earnestly besaught
this, they refused this, though assigning no
reason therefore# Aristoxenus states he heard the
this from Dionysius himself.

Hippabejus and Neanthes relate about Myllia
and Timycha.....................

ANONYMOUS

BIOGRAPHY OF PYTHAGORAS,

Preserved   by   PHOTIUS.

1. Plato was the pupil of Archytas, and thus the ninth in succession from Pythagoras; the tenth was Aristotle. Those of Pythagoras's disciples that were devoted to contemplation were called sebastici, the reverend, while those who were engaged in business were called politicians. Those who cultivated the disciplines of geometry and astronomy, were called   students. Those who associated personally with Pythagoras were called Pythagoreans, while those who merely imitated his teachings were called Pythagoristians. All these generally abstained from the flesh of animals; at a certain time they tasted #### the flesh of victims only.

2. Pythagoras is said to have lived 104 years; and Mnesarchus, one of his sons, died a young man. Telauges was another son, and Sara and Myia were his daughters. Theano, it is said, was not only his disciple, but practically his daughter.

3. The Pythagoreans preach a difference between the Monad, and the One; the Monad dwells in the intelligible realm, while the One dwells among numbers. Likewise, the Two exists among numerable things, while the Duad is indeterminate.

4. The Monad expresses equality and measure, the Duad expresses excess and defect. mean and measure cannot admit of more or less, while excess and defect, which proceed to infinity, admit it; thay is why the Duad is called indeterminate. Since, because of the all-inclusion of the Monad and Duad, all things refer to number, they call all things numbers; and number is perfected in the Ten. Ten is reached by adding in order the first four figures; that is why the Ten is called the Quaternary (or, Tetrachtys).

5. They affirm that man may improve in three ways: first, by conversation with the Gods, for to them none can approach unless he abstain from

all evil,imitating the divinity, even unto assimilation; second, by well doing, which is a characteristic of the divinity; third by dying; for
if the slight soul-separation from the body resulting from discipline improves the soul so that she
begins to divine, in dreams; and if the disease -
extasies produce visions, then the soul must surely
improve far more when entirely separated from the
body by death.

6. The Pythagoreans abstained from eating
animals, on their foolish belief in transmigration; also because this flesh-food engages diges-
tion too much, and is too fattening. Beans also
they avoided, because they produced flatulency,
produced over-satiety, and other reasons.

7. The Pythagoreans considered the Monad as the
beginning of all things, just as a point is the beginning of a line, a line of a surface, and a surface of a solid, which constitutes a body. A point
implies a preceding Monad, so that it is really
the principle of bodies, and all of them arise from
the Monad.

8. The Pythagoreans are said to have predicted
many things, and Pythagoras's predictions always
came true.

9, Plato is said to have learned his speculative and physical doctrines from the Italic Pythagoreans; and his ethics from Socrates; and his
logic from Zeno, Parmenides and the Eleatics. But
all of these teachings descended from Pythagoras.

10. According to Pythagoras, Plato and Aristotle, sight is the judge of the ten colors; white
and black being the extremes of all others, between:
yellow, tawny, pale, red, blue, green, light blue,
and grey. Hearing is the judge of the voice, sharp
and flat. Smell judges of odors, good and bad, and
putridity, humidity, liquidness and evaporation.
Taste judges of tastes, sweet and bitter, and between them five: sharp, acid, fresh, salt and hot.
Touch judges of many things between the extremes of
heavy and lightness, such as heat and cold; and thos
those between them, hardness and softness; and thos
those between them, dryness and moistness, and thos
those between them. While the main four senses are
confined to their special senses in the head, touch

IS DIFFUSED THROUGHOUT THE HEAD AND the whole body,
and is common to all the senses; but is specialised
in the hands.

11. Pythagoras taught that in heaven there were
twelve orders: the first and outermost being the
fixed sphere, where, accirding to Aristotle, dwelt
the highest God, and the intelligible deities;
and where Plato located his ideas. Next are the
seven planets: Saturn, Jupiter, Mars, Venus, Mere y
cury, Sun and Moon. Then comes the sphere of Fire,
that of Air, Water, and last, Earth. In the fixed
sphere dwells the First Cause, and whatever is   ¶
nearest thereto is the best organized, and most
excellent; while that which is furthest therefrom
is the worst. Constant order is preserved as low
as the Moon; while all things sublunary are disorderly. Evil, therefore, must necessarily exist
in the neighborhood of the Earth; which has been
arranged as thelowest, as a basis for the world,
and as a receptable for the lowest things. All   ⸶
superlunary things are governed in firm order,
and Providentially; and the decree of God, which
they follow; while beneath the moon operate four
causes: God, Fate, our election, and Fortune.
For instance, to go aboard a ship, or not, is in
our power; but the storms and tempests that may
arise out of a calm, are the result of Fortune;
and the preservation of the ship, sailing through
the waters, is in the hands of Providence, of God.
There are many different modes of Fate. There is
a distinction to be made between Fate, which is
determined, orderly and consequent, while Fortune
is spontaneuus and casual. For example, it is one
mode of Fate that guides the growth of a boy   ●
through all the sequent ages to manhood.

12. Aristotle, who was a diligent investigator,
agreed with the Pythagoreans that the Zodiac runs
obliquely, on account of the generations of those
rthly things which became complements to the
verse. For if these moved evenly, there would be
no change of seasons, of any kind. Now the paste of the sun and the other planets from one
n to another effect the four seasons of the
r, which determine the growth of plants, and
eration of animals.

13. Others thought that the sun's size exceeded that of the earth by no more than thirty times; but Pythagoras, as I think correctly, taught it was more than a hundred times as great.

14. Pythagoras called the revolution of Saturn the great year, inasmuch as the other planets run their course in a shorter time; Saturn, thirty years; Jupiter, in twelve; Mars in two; the Sun in one; Mercury and Venus the same as the Sun. The Moon, being nearest to the Earth, has the smallest cycle, that of a month.

##. It was Pythagoras who first called heaven cosmos, because it is perfect, and "adorned" with infinite beauty and living beings.

##. With Pythagoras agreed Plato and Aristotle, ✓ that the soul is immortal; although some who did not understand Aristotle claimed he taught the soul soul was mortal.

##. Pythagoras said that man was a microcosm; which means, a compendium of the universe; not becau not because, like other animals, even the least, he is constituted by the four elements, but because he contains all the powers of the world. For the world contains gods, the four elements, animals and plants. All of these powers are contained in man. He has reason, which is a divine power; he has the nayure of the elements, the powers of moving, growing, and reproduction. However, in each of these he is inferior to the others. For example, an athlete who practices five kinds of sports, and diverting his powers into five channels, is inferior ior to the athlete who practises a single sport; so man, having all of the powers, is inferior in each. Than the gods, we have less reasoning powers; and less of each of the elements than the elements themselves. Our anger and desire are inferior to these passions in the irrational animals; while our powers of nutrition and growth are inferior to that in plants. Constituted therefore of different powers, we have a difficult life to lead.

16. While all other things are ruled by one nature only, we are drawn by different powers; as for instance, when by God we are drawn to better things, or when we are drawn to evil courses by the prevailing of the lower powers. He who,

LIKE A VIGILANT AND EXPERT CHARIOTEEr, within himself cultivates the divine element, will be able to
utilize the other powers by a mingling of the elements , by anger, desire and habit, just as far
as may be necessary. Though it seems easy to
know yourself, this is the most difficult of all
things. This is said to derive from the Pythian
Apollo, though it is also attributed to Chilo, one
of the seven sages. Its message is, in any event,
to discover our own power; which amounts to learning the nature of the whole extant world, which, as
as God advises us, is impossible without philosophy.

17. There are eight organs of knowledge: sense,
imagination, art, opinion, prudence, science,
wisdom and mind. Art, prudence, science and mind
we share with the gods; sense and imagination, with
the irrational animals; while opinion alone is
our characteristic. Sense is a fallacious knowledge
derived through the body; imagination is a notion in
in the soul; art is a habit of cooperating with
reason. The words "with reason," are here added,
for even a spider operates, but it lacks reason.
prudence is a habit selective of the rightness
of planner deeds; science is a habit of those
things which remain ever the same, with Sameness;
wisdom is a knowledge of the first causes; while
mind is the principle and fountain of all good
things.

18. Docility is divided into three: shrewdness,
memory and acuteness. Memory guards the things
which ahve been learned; acuteness is quickness
of understanding, and shrewdness is the ability of
deducing the unlearned from what one has learned
to investigate.

19. Heaven may be divided into three: the
first sphere; second, the space from the fixed
sphere to the moon; third, the whole world, heaven
and earth.

20. The extreme elements, the best and the
worst, operate unintermittently. There is no inter-
mission with God, and things near him in mind and
reason; and plants are continuously nourished by
day and night. But man is not always active, nor

are irrational animals, which rest and sleep most
of the time.

21. The Greeks always surpassed the Barbarians
in manners and habits, on account of the mild
climate in which they live. The Scythians are trou-
troubled by cold, and the Aethicpóans by heat;
which determines a violent interior heat and moisture, resulting in violence and audacity. Analos
goulsy, those who live near the middle zone and
the mountains participate in the mildness of the
country they inhabit. That is why, as Plato says,
the Greeks, and especially the Athenians improved
the disciplines that they had derived from the
Barbarians.

22. (From them had come) strategyn painting,
mechanics, polemics, oratory, and physical culture. But the sciences of these were developed
by the Atheniens, owing to the favoabhie   natural
conditions of light, and purity of air, which had
the double effect of drying out the earth, as it
is in Attica, but making subtle the minds of men.
So a rarefied atmosphere is unfavorable to the
fertility of the earth, but is favorable to mbnt-
al development.

(In Photius's work, this is followed by a paragraph on the Etesian winds, which has nothing whatever to do with the subject, and which, therefore,
is omitted.)

LIFE OF PYTHAGORAS

By DIOGENES LAERTIUS

CHAPTER I

EARLY LIFE

Since we have niw gone through the Ionaian p philosophy, which was derived from Thales, and the lives of the several illustrious men who were the chief ornaments of that school, we will now proceed to treat of the Italian School, which was founded by Pythagoras, the son of Mnesarchus, a seal engraver, as he is recorded to have been by Hermippus; a native of Samos, or, as Aristoxenus asserts, a Tyrrhenian, and a native of one of the islands which the Athenians, after they had driven out the Tyrrhenians, had occupied. But some authors say that he was the son of Marmacus, the son of Hippasus, the son of Euthyphron, the son of Cleonymus, who was an exile from Phlias; and that Marmacus settled in Samos, and that from this this circumstance Pythagoras was called a Samian. After that, he migrated to Lesbos; having come to Pherecydes, with letters from his uncle Zoilus. Then he made three silver goblets, and carried them to Egypt as a present for each of the three priests. He had brothers, the eldest of whom was named Eunomus, the middle one Tyrrhenius, and a slave named Zamolxis, to whom the Getae sacrifice, believing him to be the same as Saturn, according to the accounr to the account of Herodotus (4:93).

II.   STUDIES

He was a pupil, as I have already mentioned, of Pherecydes the Syrian; and after his death he came to Samos, and became a pupil of Hermodamas, the descendant of Creophylus, who was already an old man now.

III   INITIATIONS

As he was a youth devoted to learning, he
Quitted his country, and got initiated into all the
the Grecian and barbarian sacred mysteries. Accordingly he went to Egypt, on which occasion Polycrates gave him a letter of introduction to Amasis;
and he learned the Egyptian language, as Antipho
tells us, in his treatise on those men who have
become conspicuous for virtue; and he associated
with the Chaldeans and Magi.

Afterwards he went to Crete, and in company with
Epimenides, he descended into the Idaean cave —
and in Egypt too he had entered into the holiest
parts of their temples, — and learned all the
most secret mysteries that relate to their Gods.
Then he returned again to Samos, and finding his
country reduced under the absolute dominion of
Polycrates, he set sail, and fled to Crotona in
Italy. Having given laws to the Italians, he there
gained a very high reputation, together with his
scholars, who were about three hundred in numbers,
and governed the republic in a most excellent manner; so that the constitution was very nearly an
aristocracy.

IV   TRANSMIGRATION

Heraclides Ponticus says that he was accustomed
to speak of himself in this manner: that he had
formerly been Aethalides, and had been accounted
the son of Mercury; and that Mercury had desired
him to select any gift he pleased except immortality. Accordingly, he had requested that, whether
living or dead, he might preserve the memory of w
what had happened to him. While, therefore, he was
alive, he recollected everything; and when he was
dead, he retained the same memory. At a subsequent
period he passed into Euphorbus, and was wounded
by Menelaus. While he was Euphorbus, he used to
say that he had formerly been Aethalides; and that
he had received as a gift from Mercury the perpetual transmigration of his soul; so that it was con-
stantly transmigrating and passing into whatever

PLANTS OR ANIMALS IT PLEASED; and he had also received the gift of knowing and recollecting all
that his soul had suffered in hell, and what sufferir
ferings too are endured by the rest of the souls.
   But after Euphorbus died, he said that his soul
had passed into Hermotimus; and when he wished to
convince people of this, he went into the territ-
ory of the Branchidae, and going into the temple
of Apollo, he showed his shield which Menelaus had
dedicated there as an offering. For he said that
he, when he sailed from Troy, had offered up his
shield which was already getting worn out, to Apollo, and that nothing remained but the ivory face
which was on it. He said that when Hermotimus died
he had become Pyrrhus, a fisherman of Delos; and that
tthat he still recollected everything, how he had
formerly been Aethalides, then Euphorbus, then
Hermotimus, and then Pyrrhus. When Pyrrhus died,
he became Pythagoras, and still recollected all
the circumstances I have been mentioning.

V. WORKS OF PYTHAGORAS

   Now thay say that Pythagoras did not leave
behind him a single book; but they talk foolishly;
for Heraclitus, the natural philosopher, speaks
plainly enough of him, saying, "Pythagoras, the
son of Mnesarchus, was the most learned of all
men in history; and having selected from these
writings, he thus formed his own wisdom and extensive learning, and mischievous art." Thus he speaks,
because Pythagoras, in the beginning of his treatise on natural philosophy, writes in the following
manner: "By the air which I breathe, and by the
water which I drink, I will not endure to be blamed
on account of this discourse."
   There are three volumes extant written by Pythagoras: one on education, one on politics, and one
on Natural Philosophy. The treatise which is now
extant under the name of Pythagoras is the work of
Lysis, of Tarentum, a philosopher of the Pythagorean school, who fled to Thebes, and became the' teac
teacher of Epaminondas. Heraclides, the son of
Sarapion, in his Abridgment of Sotion, says that
he wrote a poem in epic verse upon the Universe;

AND BESIDES THAT A SACRED poem which begins thus:
"Dear youths, I warn you cherish peace divine,
And in your hearts lay deep these words of mine."
A third about the Soul; a fourth on Piety; a fifth
entitled Helothales, which was the name of the father of Epicharmus of Cos; a sixth, called Crotona;
and other poems too. But the mystic discourse which
is extant under his name, they say is really the wor
work of Hippasus, having been composed with a view
to bring Pythagoras into disrepute. There were also
many other books composed by Aston# of Crotona,
and attributed to Pythagoras.

Aristoxenus asserts that Pythagoras derived the
greater part of his ethical doctrines from Themisto-
clea, the priestess at Delphi. Ion of Chios, in his
Victories, says that he wrote some poems and attributed them to Orpheus. His also, it is said, is
the poem called Scopadaea, which begins thus:
"Behave not shamelessly to any one."

VI GENERAL VIEWS ON LIFE.

Sosicrates, in his Successions, relates thatnhe
having been asked by Leon, the tyrant of the Phliasians, who he was, replied, "A philosopher." He
adds that pythagoras used to compare life to a festival. "And as some people come to the festival to
contend for the prizes, and others for the purposes
of traffic, and the best as spectators, so also in
life the men of slavish dispositions are born hunters after glory and covetousness; but philosophers
are seekers after the truth." Thus he spoke on this
subject. But in the three treatises above mentioned,
the following principles are laid down by Pythagoras:

He forbids men to pray for anything in particular for themselves, because they do not know what
is good for them. He calls drunkenness an expression identical with ruin, and rejects all superfluity, saying, "That no one ought to exceed the
proper quantity of meat and drink." On the subject
of venereal pleasures, he writes thus: "One ought
to sacrifice to Venus in the winter, not in the
summer; and in autumn and spring in a lesser degree. But the practice is pernicious at every season, and is never good for the health." And once,

WHEN HE WAS ASKED WHEN A MAN MIGHT indulge in the
pleasures of love, he replied, "Whenever you wish
to be weaker than yourself."

VII AGES OF LIFE

Thus does he divide the ages of life. A boy for t
twenty years; a young man -- neaniskos, -- for
twenty years; a middle aged man -- neanias, --
for twenty years, and an old man for twenty. These
different ages correspond proportionately to the
seasons; boyhood answers to the spring; youth to
summer; middle age to autumn; and old age to
winter. He uses neaniskos here as equivalent to
meirakion; and neanias as equivalent to aner.

VIII. SOCIAL CUSTOMS

Timaeus says that he was the first person to
assert that the property of friends i s common,
and that friendship is equality. His disciples
used to put all their possessions together into
one store, and use them in common. For five years
they kept silence, doing nothing but listening to
discourses, and never once seeing Pythagoras, until they were approved; after that time they were
admitted into his house, and allowed to see him.
They also abstained from the use of cypress coffins, because the sceptre of Jupiter was made of
that wood, as Hermippus tells us in the second
book of his account of Pythagoras.

ix, DISTINGUISHED APPEARANCE

He is said to have been a man of the most dignified appearance; and respecting him his disciples
adopted an opinion that he was Apollo who had come
from the Hyperboreans; and it is said that once
when he was stripped naked he was seen to have a
golden thigh. Many people affirmed that when he was
crossing the river Nessus, it addressed him be his
name.

#X WOMEN DEIFIED BY MARRIAGE

Timaeus, in the tenth book of his Histories
tells us that he used to say that women who were
married to men had the names of Gods, being succes-

sively called virgins, nymphs, and then mothers.

XI   SCIENTIFIC   CULTURE

Also it was Pythagoras who carried geometry
to perfection, after Moeris had first found out
the principles of the elements of that science,
as Aristiclides tells us in the second book of l.
History of Alexander; and the part of the scien:
to which Pythagoras applied himself above all o
ers, was arithmetic. He also discovered the nume.
ical relation of sounds on a single string; he a
studied medicine. Apollodorus the logician, reco.
of him that he sacrificed a hecatomb, when he ha
discovered that the square of the hypothenuse of
a right-angled triangle was equal to the squares
of the sides contsining the right angle. There i.
an epigram which is couched in the following ter.
"When the great Smian sage his nobly problem fou.
A hundred oxen with their life-blood dyed the
                                        ground."

XII.   DIET AND SACRIFICES

He is also said to have been the first man wh:
trained athletes on meat; and Eurymenes was the
first man, according to the statement of Phavorinus, in the third book of his Commentaries, wh
ever did submit to this diet, as before that tir.
man used to train themselves on dry figs, and
moist cheese, and wheaten b read; as the same pl.
vorinus informs us in the eighth book of his Un:
versal History. But some authors state that a t;.
trainer of the name of Pythagoras certainly did
train his athletes on this system, but that it
was not our philosopher; for that he even forbac:
men to kill animals at all, much less would he l:
have allowed his disciples to eat them, as havin:
a right to live in common with mankind. And this
was his pretext; but in reality he prohibited th.
eating of animals because he wished to train and
accustom men to simplicity of life; so that all
their food should be easily procurable, as it
would be, if they ate only such things as requir
ed no fire to cook them, and if they drank plain
water; for from this diet they would derive heal.
of body, and acuteness of intellect.

The only altar at which he worshipped was that
of Apollo the Father, at Delos, which is at the
back of the altar of Ceratinus, because wheat and
barley, and cheese-cakes are the only offerings
laid upon it, as it is not dressed by fire; and no
victim is evr slain there, as Aristotle tells us,
in his Constitution of the Delians. It is also said
that he was the first person who asserted that the
soul went a necessary circle being transformed and
confined at different times in different bodies.

XIII   MEASURES AND WEIGHTS

He was also the first person who introduced measures and weights among the Greeks, as Aristoxenus the musician informs us.

XIV   HESPERUS LUCIFER

Parmenides assures us too that he was the first
person who asserted the identity of Hesperus
and Lucifer.

XV STUDENTS AND REPUTATION

He was so greatly admired that it used to be said
that his disciples looked on all his sayings as
the oracles of God. In his writings he himself
said that he had come among men after having spent
two hundred and seven years in the shades below.
Therefore the Lucanians, Peucetians, Messapians
and Romans flocked around him, coming with eagerness to hear his discourses; but until the time of
Philolaus no doctrines of Pythagoras were ever
divulged; and he was the first person who published the three celebrated books which Plato wrote to
have purchased for him for a hundred minae. The
scholars who used to come to him by night were
no less than six hundred. Whenever any one of them
was permitted to see him, he wrote of it to his
friends, as if they had achieved something wonderful.

The people of Metapontum used to call his house t
the temple of Ceres; and the street leading to it
was called that of the Muses, as we are informed

in the universal history of Phavorinus.

According to the account given by Aristoxen
in his tenth book of his Laws on Education, the
rest of the Pythagoreans used to say that his pr
cepts ought not to be divulged to all the world;
and Xenophilus the Pythagorean, when he was aske
what was the best way for a man to educate his s
son, said, "That he must first of all take care
that he was born in a city which enjoyed good
laws."

Pythagoras formed many excellent men in Italy, by his precepts, and among them Zaleucus
and Charondas, the law-givers.

XVI   FRIENDSHIP FOUNDED ON SYMBOLS

Pythagoras was famous for his power of attracting friendships; and among other things, if he
ever heard that anyone had any community of symbols with him, he at once made him a companion
and a friend.

XVII   SYMBOLS OR MAXIMS

Now what he called his symbols were such as these.
"Do not poke the fire with a sword." "Do not sit
down on a bushel." "Do not devour your heart."
"Do not aid men in discarding a burden, but in
increasing one." "Always have your bed packed up."
"Do not bear the image of God on a ring." "Efface
the traces of a pot in the ashes." "Do not wipe a
seat with a lamp." "Do not make water in the sun-
shine." "Do not walk in the main street." "Do not
offer your hand lightly." "Do not cherish swallows under your roof." "Do not cherish birds with
crooked talons." "Do not defile; do not stand upo
the parings of your nails, or the cuttings of you
hair." "Avoid a sharp sword." "When travelling
abroad, do not look back at your own borders."

Now the precept not to poke the fire with
a sword meant, not to provoke the anger or swel-
ling pride of powerful men; not to violate the
beam of the balance meant, not to transgress fair-
ness and justice; not to sit on a bushel is to
have an equal care for the present and the future;

for by the bushel is meant one's daily food. By
not devouring one's heart, he intended to show
that we ought not to waste away our souls with
grief and sorrow. In the precept that a man when
travelling abroad should not turn his eyes back,
he recommended those who were departing this life
not to be desirous to live, and not to be too much
attracted by the pleasures here on earth. And the
other symbols may be explained in a similar manner,
that we may not be too prolix here.

XVIII   PERSONAL   HABITS

Above all things, he used to prohibit the eating
of the erythinus and the melanurus; also the hearts
of animals, and beans. Aristotle informs us that
to these prohibitions he sometimes added tripe
and mullet. Some authors assert that he himself
used to be contented with honey, honey-comb and
bread; and that he never drank wine in the day-
time. He usually ate vegetables, either boiled or
raw; and he very rarely ate fish. His dress was
white, very clean; his bed-clothes also were white,
and wollen, for linen had not yet been introduced
in that country. He was never known to have eaten
too much, or to have drunk too much; or to indulge
in the pleasures of love. He abstained wholly from
laughter, and from all such indulgences as jests
and idle stories. He never chastised any one, whether slave or free man, while he was angry. Admonishing he used to call feeding storks.

He used to practise divination, as far as aug-
guries and auspices; but not by means of burnt-
offerings, except only the burning of frank-incens
cense. All the sacrifices which he offered consisted of inanimate things. But some, however, assert that he did sacrifice animals, limiting himself
self to cocks, and sucking kids, which are called
apalioi, but that he very rarely offered lambs.
Aristoxenus, however, affirms that he permitted th
the eating of all other animals, and abstained
only from oxen used in agriculture, and from rams.

XIX   VARIOUS TEACHINGS

The same author tells us, as I have already men-
oned, that he received his doctrines from Themi-
clea at Delphi. Hieronymus says, that when he decended into the shades below, he saw the soul of
Hesiod bound to a brazen pillar, and gnashing it
teeth; and that of Homer suspended from a tree,
snakes around it, as a punishment for the thing
that they had said of the Gods. Those who refrai
from commerce with their wives also were punishe
and that on account of this he was greatly honor
at Crotona.

Aristippus of Cyrene, in his Account of Nat
ural Philosophers, says that Pythagoras derived
his name from the fact of his speaking (agoreu-
ein), truth no less than the God at Delphi (tou
Puthiou).

He used to admonish his disciples to repeat
these lines to themselves whenever they returned
home to their houses: --
"In what have I transgressed? What have I done?
What that I should have done have I omitted?"

He used to forbid them to offer victims to ti.
Gods, ordering them to worship only at those alt.
which were unstained with blood. He also forbad
to swear by the Gods, saying, "That every man oug
so to exercise himself as to be worthy of belief
without an oath. He also taught men that it beho-
them to honor their elders, thinking most honorat
that which was precedent inpoint of time; just as
in the world, the rising of the sun was more so t
the setting; in life, the beginning more so than
the end; and in animals, production more than des
truction.

Another of his rules was that men should hono:
the Gods above the geniuses, and heroes above men
and of all men, parents were those entitled to mo
honor. Another, that people should associate with
each other in such a way as not to make their fri
friends enemies, but to render their ~~friends~~ enem
ies friends. Another was that they should not
think anything exclusively their own. Another was
to assist the law, and to make war upon lawlessness. Not to destroy or injure a cultivated tree,

NOR ANY ANIMAL WHICH DOES not injure man. Modesty
and decorum consisted in never yielding to laughter, without looking stern. Men should avoid eating too much flesh, and in travelling should let
rest and exertion alternate; that they should exercise memory, nor ever say or do anything in anger, not pay respect to every kind of divination,
should sing songs accompanied by the lyre, and
should display a reasonable gratitude to the Gods
and eminent men by hymns.

His disciples were forbidden to eat beans, because, as they were flatulent, they greatly partook of animal properties; (that their stomachs
would be kept in much better order by avoiding
them), and that such abstinence would make the
visions that appear in one's sleep gentle and free
from agitation.

Alexander, in his Successions of Philosophers,
reports the following doctrines as contained in
Pythagoras's Commentaries: the Monad is the beginning of everything. From this proceeds an indefinite duad, which is subordinate to the monad, as to
its cause. From the monad and the indefinite duad
proceed numbers. From numbers proceed signs. From
these, lines, of which plane figures consist. From
these plane figures are derived solid bodies. From
solid bodies are derived sensible bodies, of which
last there are four elements, fire, water, earth
and air. The world, which is endued with life and
intellect, and which is of a spherical figure, in
its centre containing the earth, which is also
spherical, and inhabited all over, results from a
combination of these elements, and from them derives its motion. There are antipodes, and what
to us is below, is to them above.

He also taught that light and darkness, cold
and heat, dryness and moisture, were equally divided in the world; and that, while heat was predominant in summer, so when cold prevailed, it was
winter; when dryness prevailed, it was spring;
and when moisture preponderated, autumn. The love-
liest season of the year was when all these qualities   were equally balanced; of e -
which the flourishing spring was the most wholes-

some, and the autumn, the most pernicious. Of
day, the most flourishing period was the morn
while the evening was the fading one, and the
least healthy.

Another of his theories was that the air
around the earth was immovable, and pregnant
disease, anf that in it everything was mortal
while the upper air was in perpetual motion,
and salubrious; and that in it everything was
immortal, and on that account divine. The sun
moon and the stars were all Gods; for in them
dominates the principle which is the cause of
The moon derives its light from the sun. There
a relationship between men and the Gods, beca
men partake of the divine principle; on which
count, therefore, God exercises his providenc
for our advantage. Fate is the cause of the ar
rangement of the world, both in general and in
particular. From the sun a ray penetrates both
the cold aether, which is the air, aer, and the
dense aether, pachun aithera, which is the sea
and moisture. This ray descends into the depth
and in this way vivifies everything. Everythin
which partakes of the principle of heat lives,
which account, also, plants are animated being
but that not all living beings necessarily hav
souls. The soul is something torn off from the
aether, both warm and cold, from its partaking
of the cold aether. The soul is something diff
ent from life. It is immortal, because of the
mortality of that from which it was torn off.

Animals are born from one another by seeds,
band that it is impossible for there to be any
spontaneous production by the earth. Seed is a
drop from the brain which in itself contains a
warm vapor; and that when this is applied to
the womb, it transmits moisture, virtue, and
blood from the brain, from which flesh, sinews,
bones and hair, and the whole body are produce
From the vapor is produced the soul and also
sensation. The infant first becomes a solid be
at the end of forty days; but, according to the
principles of harmony, it is not perfect till
seven, or perhaps nine, or at most ten months,
and then it is brought forth. In itself it con-

tains all the principles of life, which are all
connected together, and by their union and combination form a harmonious whole, each of them dev-
eloping itself at the appointed time.

In general the senses, and especially sight,
are a vapor of intense heat, on which account a
man is said to see through air, or through water.
For the hot principle is opposed by the cold one;
since, if the vapor in the eyes were cold, it
would have the same temperature as the air, and
so would be dissipated. As it is, in some passa-
ges he calls the eyes the gates of the sun. In
a similar manner he speaks of hearing, and of the
other senses.

He also says that the soul of man is divided
into three parts; into intuition (nous), reason
(phren), and mind (thumas); and that the first
and last divisions are found also in other animals,
but that the middle one, reason, is found in man
only. The chief abode of the soul is in those
parts of the body which are between the heart and
the brain. The mind abides in the heart, while
the intuition (or deliberation) ############
and reason reside in the brain.

The senses are drops from them; and the
reasoning sense is immortal, while the others
are ##mortal. The soul is nourished by the blood, -
and reasons are the winds of the soul. The soul
is invisible, and so are its reasons, inasmuch as
the aether itself is invisible. The links of the
soul are the arteries, veins and nerves. When the
the soul is vigorous, and is by itself in a quies-
cent state, than its links are words and actions.
When it is cast forth upon the earth, it wanders
about, resembling the body. Mercury is the stew-
ard of the souls, and that is the reason of his
name Conductor, Commercial, and Infernal, since
it is he who conducts the souls from their bodies,
and from earth, and sea; and that he conducts the
pure souls to the highest region, and that he does
not allow the impure ones to approach them, nor to
come near one another; committing them to be bound
in indissoluble fetters by the Furies.

The Pythagoreans also assert that the whole
air is full of souls, and that these are those that

that are accounted geniuses or heroes. They are
the ones that send down among men dreams, and
tokens of disease and health; the latter not
being reserved to human beings, but being sent
also to sheep and other cattle. They are conce
ned with purifications, expiations, and all kii
of divinations, oracular predictions, and the 1

Man's most important privilege is to be abl
to persuade his soul to be either good or bad.
are happy when they have a good soul; yet they
never quiet , never long retaining the same min.
An oath is justice; and on that account Jupiter
is called Jupiter of Oaths. Virtue is harmony,
health, universal good and God; on which accoun.
everything owes its existence and preservation
to harmony. Friendship is a harmonious quality.

Honors to Gods and heroes should not be equa
The Gods should be honored at all times, extolli
them with praises, clothed in white garments, an
keeping one's body chaste; but that to the heroe
such honors should not be payed till after noon.

A state of purity is brought about by purif.
cations, washings and sprinklings; by a man's pu.
fying himself from all funerals, concubinage, or
any kind of pollution; by abstaining from all fl
that has either been killed or died of itself, f:
from mullets, from melanuri, from eggs, from suci
animals as lay eggs, from beans, and from other
things that are prohibited by those who have cha:
of the mysteries in the temples.

In his treatise on Beans, Aristotle says that
Pythagoras's reason for demanding abstention fro:
them on the part of his disciples, was that eith
they resemble parts of the human body, or becaus
they are like the gates of hell — they are the
only plants without parts; — or because they dr:
up other plants, or because they are represent-;
atives of universal nature, or because they are
used in elections in oligarchical governments.

He also forbade his disciples to pick up what
fell from the table, for the sake of accustoming
them to eat moderately, or else because such
things belong to the dead. Aristophanes, indeed,
said that what fell belonged to the heroes, in
his Heroes singing,

"   "Never taste the things which fall,
     From the table on the floor."
     He also forbade his disciples to eat white
poultry, because a cock of that color was sacred
to the God Month, and was also a suppliant. He was
also accounted a good animal (?) and he was sacred to the god Month, for he indicates the time.
     The Pythagoreans were also forbidden to eat
of all fish that was sacred, on the ground that
the same animals should not be served up before
both gods and men, just as the same things do
not belong to both freemen and slaves. Now white
is an indication of a good nature, abd black of
a bad one.
     Another of the precepts of Pythagoras was
never to break bread; because in ancient times
friends used to gather around the same loaf, as
they even now do among the barbarians. Nor would
he allow men to divide bread which unties them. Some
Some think that he laid down this rule in reference
ence to the judgment which takes place in hell;
some because this practice engenders timidity
in war. According to others, the refence is to
the Union, which presides over the government of
the Universe.
     Another one of his doctrines was that of all
solid figures the sphere was the most beautiful;
and of all plane figures, the circle. That all
age, and all diminution was similar, and also
all increase and youth. That health was the permaner
tanence of form, and disease, its destruction.
He thought salt should be set before people as a rem
reminder of justice; for salt preserves everyth
thing which it touches, and is composed of the
purest particles of water and the sea.
     These are the doctrines which Alexander asserts
that he discovered in the Pythagorean treatises;
and Aristotle gives us a similar account of them.

XX. POETIC TESTIMONEES

     Timon, in his Silli, has not left unnoticed
the lignified appearance of Pythagoras, though he
he attacks him on other points. Thus he speaks:

"Pythagoras who often teaches
Precepts of magic, and with speeches
Of long high-sounding diction draws,
From gaping crowds, a vain applause."
Referring to his having been different people at
different times, Xenophanes says in an elegiac
poem, that begins thus:
Now will I upon another subject touch,
And lead the way...........
They say that once, as passing by he saw
A dognseverely beaten, he did pity him;
And spoke as follows to the man who beat him:
"Stop now, and beat him not; since in his body
Abides the soul of a dear friend of mine,
Whose voice I recognized as he was crying."
Cratinus also ridiculed him in his Pythagorean
Woman; but in his Tarentines he speaks thus:
"They are accustomed, if by chance they see
A private individual abroad,
To try what powers of argument he has,
How he can speak and reason; and they bother him
With strange antithesis, and forced conclusions,
Errors, comparisons, and magnitudes,
Till they have filled, and quite perplexed his
                                        mind."

In his Alcmaeon, Innesimachus says:
"As we do sacrifice to the Phoebus whom
Pytagoras worships, never eating aught
Which has the breath of life."
Austophon says in his Pythagorean:
A. "He said that when he did descend below
Among the shades in Hell, he there beheld
All men who e'er had died; and there he saw,
That the Pythagoreans differed much
From all the rest; for that with them alone
Did Pluto deign to eat, much honoring
Their pious habits.
B. "He's a civip God,
Ig he likes eating with such dirty fellows."
And again in the same play he says,
"They eat
Nothing but herbs and vegetables, and drink
Pure water only; but their lice are such
Their cloaks so dirty, and their unwash'd soc
So rank, that none of our younger men
Will for a moment bear them.

XXI DEATH OF PYTHAGORAS

Pythagoras died in this manner. When he was
sitting with some of his companions in Milo's
house, some of those whom he did not think worthy
of admission into it, was by envy excited to set
fire to it. But some say that the people of Crotona themselves did this, being afraid lest he
might aspire to the tyranny. Pythagoras was
caught as he was trying to escape; and coming to
a place full of beans, he stopped there, saying
that it was better to be caught than to trample
on the beans, and better to be slain than to
speak; and so he was mursered by those who were
pursuing him. In this way also, mist of his companions were slain; being about forty in mumber;
but that a very few did escape, among whom were
Archippus of Tarentum, and Lysis, whim I have mentioned before.

But Dicaearchus ibates that Pythagoras died
later, having escaped as far as the temple of the
Muses at Metapontum, where he died of starvation,
after forty days. Heraclides, in his abridgment,
of the life of Satyrus says that after he had
buried Pherecydes at Delos, he returned to Italy,
and there finding a superb banquet prepared at
the house of Milo, of Crotona, he left that city
for Metapontum, where, not wishing any longer to
live, he put an end to his life by starvation.
But Hermippus says that when there was war between
the Agrigentines and the Syracusans, Pythagoras,
with his usual companions, joined the Agrigent-
ine army, which was put to flight. Coming up ag-

ainst a field of beans, instead of crossing it,
he ran around it, and so was slain by the Syracusans; and that the rst, about thirty-five in
number, were burned at Tarentum, where they
were trying to excite a sedition in the state
against the principal magistrates.

Hermippus also relates another story about Pythagoras. When in Italy, he made a subterranean
apartment, and charged his mother to write an
account of everything that took place, marking
the time of each on a tablet, then sending them down
down to him until he came up again. His mother

did so. Then after a certain time Pythagoras came
up again, lean, and reduced to a skeleton; he
came into the public assembly, and said that he
had arrived from the shades below, and then he
recited to them all that had happened to them
in the meanwhile. Being charmed with what he told
them, they believed that Pythagoras was a divine
being, so they wept and lamented, and even entrusted to him their wives, as likely to learn some
good from him; and they took upon themselves the
name of Pythagoreans. Thus far Hermippus.

XXII   PYTHAGORAS'S   FAMILY

Pythagoras had a wife, whose name was T heane, the
daughter of Brentinus of Crotona. Some say that she
she was the wife of Brontinus, and only Pythagoras's pupil. As Lysis mentions in his letternto
Hipparchus, he had a daughter named Damo. Lysis's
letter speaks of Pythagoras thus: "And many say
that you philosophize in public, as Pythagoras
also used to do; who, when he had entrusted his
commentaries to his daughter Damo, charged her
not to divulge them to any one outside of the
house. Though she might have sold his discourses
for much money, she did not abandon them; for she
thought that obedience to her father's injunctions,
even though this entailed poverty, better than
gold; and that too, though she was a woman."
   He had also a son, named Telauges, who was his
father's successor in his school; and who, according to some authors, was the teacher of Empedocles.
At least Hippobotus relates that Empedocles said,
   "Telauges, noble youth, whom in due time
   Theano bore, to wise Pythagoras."
   But there is no book extant, which is the
work of Telauges, though thereare some extant
that are attributed to his mother Theano.Of her
is told a story, that once, when asked how long a
woman should be absent from her husband, and remain
ain pure, she said, The moment she leaves her own
husband, she is pure; but she is never pure at all,
after she leaves any one else. A women who was
going to her husband was by her told to put off her
modesty with.her clothes, and when she left him,

to resume it with her clothes; when she was asked,
What clothes, she said, Those which cause you to bo
be called a woman."

XXIII   RIDICULING   EPIGRAMS

Now Pythagoras, according to Heraclides, the so
son of Serapion, died when he was eighty years of
age, according to his own account; by that of oth-
ers, he was over ninety. On him we have written
a sportive epigram, as follows:
"You are not the only man who has abstained
From living food; for so have we;
And who, I'ld like to know, did ever taste
Food while alive, most sage Pythagoras?
When meat is boiled, or roasted well and salted,
I do not think it well can be called living.
Which, without scruple therefore then we eat it,
And call it no more living flesh, but meat."
Another, which runs thus:
"Pythagoras was so wise a man, that he
Never ate meat himself, and called it sin.
Yet gave he good joints of beef to others;
So that I marvel at his principles;
Who others wronged, by teaching them to do
What he believed unholy for himself."
Another, which follows:
"Should you Pythagoras's doctrine wish to know,
Look on the centre of Euphorbus's shield.
For he asserts there lived a man of old,
And when he had no longer an existence,
He still could say that he had been alive,
Or else he would not still be living now."
Nother one follows:
"Alas!   alas! why did Pythagoras hold
Beans in such wondrous honor? Why, besides,
Did he thus die among his choice companions?
There was a field of beans; and so the sage,
Died in the common road of Agrigentum,
Rather than trample down his favorite beans.

XXIV   THE LAST PYTHAGOREANS

He flourished about the sixtieth olympiad; and
his system laste for about nine or ten generations.

The last Pythagoreans known to Aristoxenus were
Xenophilus the Chalcidean, from Thrace; Phanton
the Phliasianwith his countrymen Echurates, Diod.
and Polymnestus, disciples of Philolaus and Eu-·
rytus of Tarentum.

XXV   VARIOUS PYTHAGORASES

Pythagoras was the name of four men, almost con-
temporaneous, and living close to each other. One
was a native of Crotona, a man who attained to
tyrant's power; the second was a Phliasian, and
as some say, a trainer of wrestlers. The third
was a native of Zacynthus; the fourth was this
our philosopher, to whom the mysteries of philosophy are said to belong, and in whose time the
proverbial phrase, ipse dixit, arose generally.
Some also claim the existence of a fifth Pythagoras, a sculptir of Rhodes, who is believed to
have been the first discoverer of rhythm and proportion. Another was a Samian sculptor. Another,
an orator of small reputation. Another was a phys
ician, who wrote a treatise on squills, and some
essays on Homer. Dionysius tells us there was another who wrote a history of the affairs of the
Dorians.

Eratosthenes, quoted by Phavorinus, in the eighth book of his Universal History, tells us that
this philosopher, of whom we are speaking, was the
first man who ever practised boxing in a scientific manner, in the forty-eighth olympiad, having hi
hair long, and being robed in purple. From compet-
ition with boys he was rejected; but being ridiculed for his application for this, he immediately
entered among the men, and was victorious. Among o
other things, this statement is confimed by an
epigram of Theaetetus;
  "Stranger, if e'er you knew Pythagoras,
  Pythagoras, the man with flowing hair,
  The celebrated boxer, erst from Samos,
  I am Pythagoras. And if you ask
  A citizen of Elis of my deeds,
  You will surely think he is relating fables.
    Phavorinus says that he employed definitions,
  on account of the mathematical subjects to which

he applied himself. Socrates and his pupils did
so still more; and in this they were later followed by Aristotle and the Stoics.

He too was the first man who applied to the
universe the name kosmos, and who first called the
earth round; though Theophrastus attributes this
to Parmenides, and Zeno to Hesiod. It is also
said that he had a constant adversary, named Cylon,
as Socrates's was Antidicus. This epigram was formerly repeated concerning Pythagoras the athlete:
"Pythagoras of Samos, son of Crates,
Came while a child to the Olympic games;
Eager to battle for the prize in boxing."

XXVI PYTHAGORAS'S LETTER

Extant is a letter of our philosopher's, which
follows:
PYTHAGORAS TO ANAXIMENES
"YOU TOO, most excellent friend, if you were
not superior to Pythagoras in birth and reputation, would have migrated from Miletus, and gone
elsewhere. But now the reputation of your father
keeps you back, which perhaps would have rest
trained me too, if I had been like Anaximenes.
But if you, who are the most eminent man, abandon the cities, all their ornaments will disappear, and the Median power will be the more dangerous to them. Nor is it always seasonable to
be studying astronomy, but it is more honorable
to exhibit a regard for one's country. I myself
am not always occupied about speculations of
my own fancy, but I am busied also with the wars
which the Italians are waging one with another."

But since we have now finished our account of
of Pythagoras, we must also speak of the most
eminent of the Pythagoreans. After whom, we must
mention those who are spoken of more promiscuously
in connection with no particular school; and then
we will connect the whole series of philosophers
worth speaking of, till we arrive at Epicurus.
Now Jelanges and Theano we have mentioned; and
we must speak of Empedocles, in the first place,
for according to some accounts, he was a pupil
of Pythagoras.

EMPEDOCLES AS PYTHAGOREAN

Timaeus, in his ninth book, relates that he was
a pupil of Pythagoras, saying that he was afterwards convi cted of having divulged his doctrines,
in the same way as Plato was, and that he was
therefore hencefoꞁth forbidden from attending
his school. It is said Pythagoras had him in mind
when he said:
"And in that band there was a learned men
 Of wondrous wisdom; one who of all men
 Had the profoundest wealth of intellect."
But some say the philosopher was here referring
to Pythagides. PARMENIDES

  Neanthes relates that until the time of Philolaus and Empedocles the Pythagoreans used to admit into their schoold all persons indiscriminately into their school; but when Empedocles, by
means of his poems, then they made a law to admit no epic poet. They said that the same thing
happened to Plato; for that he too was excluded
from the school. Who was Empedocles's Pythagorean
teacher is not mentioned; for, as for the letter
of Jelanges, in which he is stated to have been
a pupil of Hippasus and Brontinus, that is not
worthy of belief. But Theophrastus says that he was
was an imitator and rival of Parmenides in his
poems, for that he too had delivered his opinions
on natural philosophy in Epic verse.

  Hermippus however says that he was an imitator
not of Parmenides, but of Xenophanes with whom
he lived; and that he imitated his epic style,
and that it was at a later period that he fell in
with the Pythagoreans. But Alcimadas, in his Natural Philosophy, says that Zeno and Empedocles
were pupils of Parmenides, about the same time;
and that they subsequently seceded from him. Zeno
was said to have adopted a philosophical system
peculiar to himself; but that Empedocles became
a pupil of Anaxagoras and Pythagoras, and that he it
imitated the pompous demeanor and way of life and
gestures of the one, and the system of Natural Philosophy of the other.

P.Y T H A G O R E A N L I B R A R Y

A COMPLETE COLLECTION OF

THE WORKS OF SURVIVING WORKS OF PYTHAGOREANS

For the Lack of which
Pythagoreanism has till now
Remained Unrecognized as
the Source of Greek Thought.

Translated and Collected   by

KENNETH SYLVAN GUTHRIE

Translator of Plotinus,
Compiler of the Lives of Pythagoras

T H E P L A T O N I S T P R E S S

Box 42, A L P I N E N.J.

INDEX OF PYTHAGOREAN FRAGMENTS

INTRODUCTION

The reason that Pythagoreanism has been neglected, and often treated mythically, is that until this edition the Pythagorean fragments have
never been collected, in text, or any translation.
This book therefore marks an era in the study of
philosophy, and is needed by every university and
general library in the world, not to mention those
of the students of philosophy.

But there is yet a wider group of people who
will welcome it, the lovers of truth in general,
who will be charmed by Hierocles' modern views
about the family, inspired by Iamblichus's beautiful life of Pythagoras, which has been inaccessible for over a century, and strengthened
by the maxims of Sextus, which represent the rel-
igious facts of the religion of the future more
perfectly than can easily be found elsewhere.

The universal culture of Pythagoras is faithfully portrayed by the manifold aspects of the
teachings of Archytas, and Philolaus, and of many
other Pythagoreans, among whose fragments we find
dissertations on every possible subject: metaphysics, psychology, ethics, sociology, science, and
art. Men of general culture, therefore, will feel
the need of this encyclopedic information and
study; and conversely, there is neither scient-
ist, metaphysicaian, clergyman, litterateur or
sociologist who will fail to discover therein
something to his taste.

The Fragments have been gathered from various
sources. On Philolaus, the authority is Boeckh.
The Archytas fragments have been taken from Chaignet; the minor works from Gale and Taylor,
and the Maxims and Golden verses from Dacier.
The Timaeus was taken from Plato's works, among
which it has been preserved. Hierocles's Commentary on the Golden verses has been temporarily omit
omitted as late, wordy, and containing nothing
new.

As it is the Editor's purpose to live up to
the title of this book, "A Complete Pythagorean
Library," he will be grateful to any purchaser
of the book who may point out to him further
fragments that might be added, as the Editor
has no idea that he has, in spite of his good
intentions, and herculean labors, done more than
to make the first attempt in a most important
direction. Moreover, as the work had to be done
at off times, by night, or on holidays, it was
inevitably hurried, and therefore inevitably
imperfect; for all of which oversights and errors
he begs consideration, forgiveness, and constructive criticism.

This work was done, however, because of its
great significance in the history of philosophy,
which has been elsewhere more definitely been
pointed out, and for the sake of which, no doubt,
the book will be procured by all students, philosophers, and general lovers of truth. It was undertaken for no purpose other than the benefit
of humanity1 that had for so long been deprived
of this its precious heritage, and the Editor
will be satisfied if he succeeds in restoring
these treasures of thought and inspiration to his
day and generation.

IMPORTANCE OF THIS COLLECTION

OF PYTHAGOREAN FRAGMENTS

It is a general notion among the uneducated
that the great geniuses of thought and poetry
arose by divine decree in ready-made original-
ity. Goethe did his best to disabuse the world
of this, acknowledging that most of the merit
of his work was due to the literature he had
studied better than anybody else of his circle.
Virgil was so ashamed of his borrowings from
Ennius and others, later demonstrated by Mac-
robius, that on his deathbed he wished to
destroy his Aeneid, not understanding that it
was all the more precious to us for the fidelity with which it represented the then immediately preceding age. The uncoverers of the
sources of Shakespeare, Homer, Milton (Vondel),
Dante (Bruno Latini), and many ethnic script-
ures have done their victims no harm, but rather
honor, enriching their significance,n and making
them all the more precious to the world which
in the last analysis cares nothing for a British
poacher and pawnbroker who wrote his name in 600
different ways, or about a blind traveller, compelled to make the most of his foreign findings,
or a Florentine Bolshevik exile and sycophant,
to whom it was heaven to be guided by a stout
mother of a great family, who had repulsed him;
but the world is very much concerned in having,
in modern, accessible and cheap form a summary
of the best that has been done up to that time.

In restoring the back-ground of philosophy
and thought behind Plato and Aristotle, we are
not doing them an injury, but rather making their
utterances all the more precious by showing the
mental associations that inspired them as they
peened their immortal words.

This can, of course, be done only very partially, for we have only fragments to deal with;
but the inference is reasonable that if we can

suggest so much from mere fragments, we could do
much more from the now lost complete works of
the Pythagoreans.

    To begin with, Plato showed his good taste
by making great efforts to procure the inaccessible writings of Ocellus, and through Archytas
secured several. So we have a definite historical
connection on which to base our further supposi-
tions.

    Then we hear that he paid a large sum of money
for a Pythagorean writing, which indeed may have
been the treatise of the Locrian Timaeus, which
is generally printed with his works, and whose
close relations with his own "Timaeus" are un-
blinkable. To begin with, we do know that the
titles of many of his dialogues were not taken
on chance, but represented famous thinkers in
that field, such as the Protagoras, and others.

    The correspondences between his Timaeus and
the Locrian work are so marked, that inevitably
some connection has been assumed, and in view of
Plato's fame and the Locrian's rusticity, has
generally led to calling the Locrian work an abstract of Plato's.

    But even they who stated and assumed this had
qualms of conscience. Both De Gelder and Tene-
-mann had pointed out that the Locrian "origin of
the human scul is more clearly explained" than the
Platonic; and Burges adds, that in view of this it
is hard to understand how the former could have
been an abridgment of the latter." De Gelder had
already pointed out important discrepancies, so
that the abstract theory is unstaisfactory. The
Locrian calculation from 384 (instead of Plato's 192)
through all the numbers of the scale to a total
of 114,695 is no easy matter, and impossible for
a student abstracter; this implied great mathematical and musical skill, and could not have been
made wothcut very clear purposes, which indeed
here are unmistakably Pythagorean.

    In comparing the Locrian and Platonic essays
we find the Locrian much shorter, logical, and
without any padding. It is therefore, antecedently,
much more likely to have been the source of inspiration.

Thomas Taylor had already done much in this field, which deserves, and no doubt in the future will attract serious attention. We can here mention only a few of the better known correspondences.

The second chapter of Ocellus Lucanus's treatise, is practically reproduced by Aristotle in his essay on Generation and Corruption, especially the three things necessary to generation; also the four powers, and details about matter. Several paragraphs about the mixture of the elements are taken entire. Also the expression, "as is proper, from such things as are proper, and when it is proper."

Hippodamus's mingling of democracy, aristocracy and monarchy is found in Plato's Laws, and his Statesman.

Ecphantus said that any man who has a divine conception of things is in reality a king. Plato in his Staesman said that "we must call royal him who possesses the royal science, whether or not he governs."

Callicratidas defined God as an intellectual, and incorruptible animal, while in the 12th book of his Metaphysics Aristotle says that God is an animal eternal and most excellent."

Strange to say, Plato's mother was named Pericthyone, whose namesake was one of the Pythagoreans' female philosophers. She said that those who are unfaithful to their parents must expect punishment in hell, while Olympiodorus, on the Phaedo of Plato states that the soul is by the divinity not punished through anger, but medicinally, as was implied by Pericthyone.

Aristoxenus's second paragraph is qouted in extenso in Plato's Laws, (viii, p.187,188, Bipont)

Pempelus's fragment on Parents is also quoted by Plato in the same work.

Archytas's treatment of happiness is reproduced in part in Aristotle's Nicomachean Ethics.

This most interesting topic should furnish the subject of a most valuable treatise, which will be necessary to the proper appreciation of all Greek philosophy. Who will have time for it?

PYTHAGOREAN SYMBOLS, or MAXIMS

(From Hierocles.)

1. Go not beyond the balance. (Transgress not Jus-
   Justice).
2. Sit not down on the bushel. (Do not loaf on
   your job).
3. Tear not to pieces the crown. (Do not be a
   joy-killer).
4. Eat not the heart. (Do not grieve over-
   much).
5. Do not poke the fire with a sword. (Do not
   further inflame the quarrelsome).
6. Having arrived at the frontiers, turn not
   back. (Do not wish to live over your life).
7. Go not by the public way. (Go not the broad
   popular way, that leads to destruction).
8. Suffer no swallows around your house. (Receive no swallows into your family).
9. Wear not the image of God on your ring. (Profane not the name of God).
10. Do not unload people, but load them up. (En-
    courage not idleness, but virtue).
11. Not easily shake hands with a man. (Make no
    ill-considered friendship.
12. Leave not the least mark of the pot on the
    ashes. (After reconciliation, forget the
    disagreement).
13. Sow mallows, but never eat them. Use mildness
    to others, but not to yourself.
14. Wipe not out the place of the torch. (Let not
    all the lights of reason be extinguished).
15. Wear not a narrow ring. (Seek freedom, avoid
    slavery).
16. Feed not the animals that have crooked claws.
    To your family admit no thief or traitor).
17. Abstain from beans. (Avoid farinaceous food
    causing flatulence; avoid democratic voting).
18. Eat not fish whose tails are black. (Frequent
    not the company of men without reputation).
19. Never eat the gurnet. (Avoid revenge).

20. Eat not the womb of animals. (Avoid what
    leads to generation, to love affections).

21. Abstain from flesh of animals that die of themselves. (Avoid decayed food).
22. Abstain from eating animals. (Have no conversation with unreasonable men).
23. Always put salt on the table. (Always use the principle of Justice to settle problems).
24. Never break the bread. (When giving charity, do not pare too close).
25. Do not spill oil upon the seat. (Do not flatter princes, praise God only).
26. Put not meat in a foul vessel. (Do not give good precepts to a vicious soul.
27. Feed the cock, but sacrifice him not; for he is sacred to the sun and the moon. (Cherish people who warn you, sacrifice them not to resentment).
28. Break not the teeth. Do not revile bitterly; do not be sarcastic).
29. Keep far from you the vinegar-cruet. Avoid malice and sarcasm.
30. Spit upon the parings of your nails, and on the clippings of your hair. (Abhor desires).
31. Do not urinate against the sun. (Be modest).
32. Speak not in the face of the sun. (Make not public the thoughts of your heart.
33. Do not sleep at noon. (Do not continue in darkness).
34. Stir up the bed as soon as you are risen, do not leave in it any print of the body. (When working, hanker not for luxurious ease).
35. Never sing without harp-accompaniment. (Make of life a whole).
36. Always keep your things packed up. (Always be prepared for all emergencies).
37. Quit not your post without your general's order. (Do not suicide).
38. Cut not wood on the public road. (Never turn to private use what belongs to the public).
39. Roast not what is boiled. (Never take in ill part what is done in simplicity and ignorance)
40. Avoid the two-edged sword. (Have no conversation with slanderers.

41. Pick not up what is fallen from the table.
(Always leave something for charity).
42. Abstain even from a cypress chest. (Avoid
going to funerals).
43. To the celestial gods sacrifice an odd number, but to the infernal, an even. (To God consecrate the indivisible soul, the body to hell.
44. Offer not to the gods the wine of an unpruned
vine.(Agriculture is a great piece of piety.)
45. Never sacrifice without meal. (Encourage agriculture, offer bloodless offerings.)
46. Adore the gods, and sacrifice bare-foot.
(Pray and sacrifice in humility of heart).
47. Turn round when you worship. (Adore the immen-
sity of God, who fills the universe.
48. Sit down when you worship. (Never worship in
a hurry.
49. Pare not your nails during the sacrifice.
(In the temple behave respectfully).
50. When it thunders, touch the ground. (Appease
God by humility).
51. Do not primp by torch-light. (Look at things
in the light of God).
52. One, Two. (God and Nature; all things are known
known in God).
53. Honor marks of dignity, the Throne, and the
Ternary. (Worship magistrates, Kings, Heroes,
Geniuses and God).
54. When the winds blow, adore echo. (During
revolts, flee to deserts).
55. Eat not in the chariot. (Eat not in the midst
of hurried, important business).
56. Put on your right shoe first, and wash your
left foot first. (Prefer an active life, to one of
of ease and pleasure).
57. Eat not the brain. (Wear not out the brain;
refresh yourself).
58. Plant not the Palm-tree. (Do nothing but what
is good and useful).
59. Make thy libations to the gods by the ear.
(Beautify thy worship by music).
60. Never catch the cuttle-fish. (Undertake no
dark, intricate affairs, that will wound you).

61. Stop not at the threshold. (Be not wavering, but choose your side.

62. Give way to a flock that goes by. (Oppose not the multitude).

63. Avoid the weasel. (Avoid tale-tellers).

64. Refuse the weapons a woman offers you. (Reject all suggestions revenge inspires).

65. Kill not the serpent that chances to fall within your walls. (Harm no enemy who becomes your guest or suppliant).

66. It is a crime to throw stones into fountains. (It is a crime to persecute good men).

67. Feed not yourself with your left hand. (Support yourself by honest toil, not robbery).

68. It is a horrible crime to wipe off the sweat with iron. (It is a criminal to deprive a man by force of what he earned by labor).

69. Stick not iron into the footsteps of a man. (Mangle not the memory of a man).

70. Sleep not on a grave. (Live not in idleness on the parents' inherited estates).

71. Lay not the whole faggot on the fire. (Live thriftily, spend not all at once.)

72. Leap not from the chariot with your feet close together. (Do nothing inconsiderately).

73. Threaten not the stars. (Be not angry with your superiors).

74. Place not the candle against the wall. (Persist not in enlightening the stupid).

75. Write not in the snow. (Trust not your precepts to persons of an inconstant character).

PYTHAGORAS'S GOLDEN VERSES

1 First honor the immortal Gods, as the law
                demands;
2 Then reverence thy oath, and then the illustrieu
                ous heroes;
3 Then venerate the divinities under the earth,
                due rites performing.
4 Then honor your parents, and all of your kindred
                dred;
5 Among others make the most virtuous thy friend!

6 Love to make use of soft speeches, but deeds
                that are useful;
7 Alienate not the beloved comrade for trifling
                offences,
8 Bear all you can, what you can, and you should
                be are so near to each other.
9 Take this well to heart: you must gain control
                of your habits;
10 First over stomach, then sleep, and then luxury,

11 And anger; what brings you shame, do not unto
                others;
12 Nor by yourself; highest of duties is honor
                of self.
13 Let justice be practiced in words as in deeds;

14 Then make the habit, never inconsiderately to
                act;
15 Neither forget that death is appointed to all;

16 That possessions here gladly gathered, there
                must be left;
17 Whatever sorrow the fate of the gods may
                here send us,
18 Bear, whatever may strike you, with patience
                unmurmuring.

19 To relieve it, so far as you can, is permitted;
          but reflect:
20 Not much good has Fate given to the good.

21 The speech of the people is various, now good,
          and now evil;
22 So let them not frighten you, nor keep you from
          your purpose.
23 If false calumnies come to your ears, support
          it in patience;
24 Yet that which I now am declaring, fulfil it
          full faithfully:
25 Let no one with speech or with deeds e'er
          deceive you;
26 To do or to say what is not the best.

27 Think, ere you act, that nothing stupid result;

28 To act inconsiderately is part of a fool;

29 Yet whatever later will not bring you repentance,
     ance, that you should carry through.
30 Do nothing beyond what you know; yet learn

31 What you may need; thus shall your life grow
          happy.
32 Neither grow anxious about the health of the
          body;
33 Keep measure in eating and drinking, and every
          exercise of the body;
34 By measue, I mean what later will not induce
          pain;
35 Follow clean habits of life, but not the lu-
          xurious;
36 Avoid what envy arouses,

37 At the wrong time, never be prodigal, as if
     you did not know what was proper;
38 Nor show yourself stingy; that which is medium
          is ever the best.
39 Never let slumber approach thy wearied eye-lids,

40 Ere thrice you review what this day you did:

42 Wherein have I sinned? What did I? What duty
      is neglected?
43 All, from the first to the last, review; and
      if you have erred,
44 Grieve in your spirit, rejoicing for all that
      was good.
45 With zeal and with industry, this, then, repeat;
      and learn to repeat it with joy.
46 Thus wilt thou tread on the paths of heavenly
      virtue,
47 Surely, I swear it by him who into our souls
      placed the Four (elements),
48 Him who is spring of Nature eternal. —
      Now start on your task
49 After you have implored the blessing of the
      Gods. — If this you hold fast,
50 Soon will you recognize of Gods and mortal
      men
51 The peculiar existence, how everything passes
      and returns.
52 Then will you see what is true, how Nature
      in all is most equal,
53 So that you hope not for what has no hope,
      nor that aught should escape you.
54 Men shall you find whose sorrows themselves ha
      have created,
55 Wretches who see not the Good. that is too near,
      near,
56 Nothing they hear; few know how to help them-
      selves in misfortune.
57 That is the Fate, that blinds humanity, in circl.
      circles,
58 Hither and yon, they run, in endless sorrows;

59 For they are followed by a grim companion,
      disunion within themselves,
60 Unnoticed; ne'er rouse him, and fly from
      before him!
61 Father Zeus, O free them all from sufferings
      so great,
62 Or show unto each the Genius, who is their
      guide!

— — — — — —

63 Yet, do not fear, for the mortals are divine by rac

64 To whom holy Nature everything will reveal and der
demonstrate;
65 Whereof if you have received, so keep what I
teach you;
66 For I will heal you, and you shall remain in
sured from manifold evil.
67
67 Avoid foods forbidden, reflect, that this con
tributes to the cleanliness
68 And redemption of your soul; This all, Oh, con
sider;
69 Let reason, the gift divine, be thy highest
guide;
70 Then should you be separated from the body,
and soar in the spiritual aether,
71 Then will you be imperishable, a divinity, no
longer a human!

FRAGMENTS OF
PHILOLAUS

Biography of PHILOLAUS,

By DIOGENES LAERTES

Philolaus of Crotona, a Pythagorean, was he
from whom Plato, in one of his Letters, begged
Dio to purchase Pythagorean books. He died under
the accusation of having had designs on the tyranny. I have made about him the following epigram:
"I advise everybody to take good care to avoud
suspicion; even if you are not guilty, but seem
so, you are ruined. That is why Crotona, the home-
land of Philolaus, destroyed him, because he was
suspected of wishing to establish autocracy."

He teaches that all things are produced by necessity and harmony, and he is the first who said
that the earth has a circular movement; others
however insist this was due to Hicetas of Syracuse.
He had written a single book which the philosopher P
Plato, visiting Dionysius in Sicily, bought, accordin
cording to Hermippus, from Philolaus's parents,
for the sum of 40 Alexandrian minae, whence he
drew his Timaeus. Others state that he received them a
it as a present for having obtained the liberty
of one of Philolaus's disciples, whom Dionysius had
imprisoned. In his Homonyms, Demetrius claims that
he is the first of the Pythagorean phil osophers
who made a work on nature public property. This
book begins as follows: "The world's being is the
harmonious compound of infinite and finite principles; such is the totality of the world and all
it contains."

FRAGMENTS OF PHILOLAUS,

From Boeckh.

1. (Stob.21.7; Diog.#.8.85) The world's nature is a harmonious compound of infinite and finite elements; similar is the totality of the world in itself, and of all it contains.

b. All beings are necessarily finite or infinite, or simultaneously finite and infinite; but they could not all be infinite only.

2. Now, since it is clear that the beings cannot be formed neither of elements that are all infinite, it is evident that the world in its totality, and its included beings are a harmonious compound of finite and infinite elements. That can be seen in works of art. Those that are composed of finite elements, are finite themselves; those that are composed of both finite and infinite elements, are both finite and infinite; and those composed of infinite elements, are infinite.

2. All things, at least those we know, contain number ; for it is evident that nothing whatever can either be thought or known, without number. Number has two distinct kinds: the odd, and the even, and a third, derived from a mingling of the other two kinds, the even-odd. Each of its subspecies is susceptible of many very numerous varieties; which each manifests individually.

3. The harmony is generally the result of contraries; for it is the unity of multiplicity, and the agreement of discordances.(Nicom.Arith.2:509).

4. This is the state of affairs about nature and harmony. The essence of things is eternal; it is a unique and divine nature, the knowledge of which does not belong to man. Still it would not not be possible that any of the things that are, and are known by us, should arrive to our knowledge, if this essence was not the internal foundation of the principles of which the world was founded, that is, of the finite and infinite elements. Now since these principles are not mutually similar, neither of similar nature, it would be impossible that the order of the world should

have been formed by them, unless the harmony had
intervened, in any manner whatever. Of course,
the things that were similar, and of similar nature
ure, did not need the harmony; but the dissimilar
ar things, which have neither a similar nature,
nor an equivalent function, must be organized by
the harmony, if they are to take their place in
the connected totality of the world.

5. The extent of the harmony is a fourth,
plus a fifth. The fifth is greater than the fourth
by nine eighths; for from the lowest string to
the second lowest, there is a fourth; and from
this to the next, a fifth; but from this to the
next, or "third," a fourth; and from this "third"
to the lowest, a fifth. The interval between the
second lowest and the "third" (from the top) is
nine eighths; the interval of the fourth, is
four thirds; that of the fifth, three halves;
that of the octave, the double relation. Thus
the harmony contains five nine-eighths plus two
sharps; the fifth, three nine eighths, plus one
sharp; tha fourth two nine-eighths, plus one

6. (Boethius, Music, 3:5). Nevertheless the
Pythagorean Philolaus has tried to divide the
tone otherwise; his tone's starting-point is is
the first uneven number which forms a cube, and
you know that the first uneven number was an object of veneration among these Pythagoreans. Now
the first odd number is threeq; thrice three are
nine, and nine times three is 27, which differs
from the the number 24 by the interval of one
tone, and differs from it by this very number 3.
Indeed, 3 is one eighth of 24, and this eighth p
part of 24, added to 24 itself, produces 26, the
cube of 3. Philolaus divides this number 27 in
two parts, the one greater than half, which he
calls "apotome;" the other one smaller than half,
he calls sharp; but which latterly has become
known as minor half-tone. He supposes that this
sharp contains thirteen unities, because 13 is the
difference between 256 and 243, and that this
same number is the sum of 9, 3, and unity, in which
the unity plays the part of the period, 3 of the
first odd line, and 9 of the first odd square.

18. I insist that virtue is sufficient to
preclude unhappiness, that badness precludes
happiness, if we know how properly to judge
of the genuine condition of the soul in these
two conditions; for the evil is necessarily always unhappy, whether in abundance, — which he
n does not know how properly to judge use, — or
in poverty; just as a blind man is always whether he is in brilliant light, or in darkness. But
the worthy man is not always happy; for happiness
does not consist in the possession of virtue,
but its use; just as a man who sees does not
see all the time; he will not see without light.

Life is as it were divided into two roads;
the rougher one, followed by patient Ulysses,
and the more agreeable one followed by Nestor; I
mean that virtue desires the one, but can also
follow the other. But nature cries aloud that
happiness is life desirable in itself, whose
state is assured, because one can realize one's
purposes in it, so that if life is traversed
by things one has not desired, one is not happy,
without however being absolutely unhappy. Therefore be not so bold as to insist that the worthy
man is exempt from sickness, and suffering; dare
not to say that he does not know pain; for if the
body is allowed some causes of pain, the soul sho
should also be allowed some. The griefs of the
insane lack reason and measure; while those of
the wise are contained within the measure which
reason gives to everything; but this so advertised
insensibility enervates the character of generosity of virtue, when it stands trials, great
sorrows, when it is exposed to death, suffering,
and poverty; for it is easy to support small
sorrows. You must therefore practice the "metrio-
pathy," or sorrow-standardization; so as to avoid
the insensibility just as much as the over-sens-
ibility to pain, and not in words to boast about
our strength above the measure of our human
nature.

19. We might define philosophy as the desire
of knowing and understanding things in themselves,
joined with practical virtue, inspired and realized by the love of science.

The beginning of philosophy is the science of
nature; the middle, practical life; and the end,
science itself. It is fortunate to have been well
born, to have received a good education, to have
been accustomed to obey a just rule, and to have
habits conformable to nature. One must also have
been exercised in virtue, and have been educated
by wise parents, governors and masters. It is
fine to impose the rule of duty on one self, to
have no need of constraint, to be docile to those
who give us good advice about life and science.
For a fortunate disposition of nature, and a good
education are often more powerful than lessons
to bring us to the good; its only lack would be
the efficacious light of reason, which science
gives us. Two rival directions of life contend
for mastery; practical and philosophical life.
By far the most perfect life unites them both,
and in each different path adapts itself to circumstances. We are born for rational activity;
which we call practical. Practical reason leads
us to politics; the theoretical reason, to the
contemplation of the universality of things.
Mind itself, which is universal, embraces these
two powers necessary to happiness, which we define
as the activity of virtue in prosperity; it is not
exclusively either a practical life which would
exclude science, nor a speculative life which would
exclude the practical. Perfect reason inclines
towards these two omnipotent principles, for which
man is born; the principle of society and science;
for if these opposite principles seem mutually
to interfere in their development, the political
principles turning us away from politics, and
the speculative principles turning us from specu-
lation, to persuade us to live at rest, nevertheless
nature, uniting the ends of these two movements,
shows them fused; for virtues are not contradictory and antipathetic mutually; than the harmony
of virtues no harmony is more consonant. If, from
his youth, man has subjected himself to the principles of virtues, and to the divine law of the
world harmony, he will lead an easy life; and if,
by his own inclination, he inclines towards evil,
and has the luck of meeting better guides, he will,

12. (Plut.Plac.Phil.2:5). Philolaus explains destruction by two causes; one is the fire which descends from heaven, the other is the water of the moon, which is driven away therefrom by the circulation of air; the loss of these two stars nourish the world.

13. (Diog.Laert.8:85). Philolaus was the first who said that the world moves in a circle; others attribute it to Hicatas of Syracuse.

b.Plut.Plac.Philos.3:7). Some insist that the earth is immovable; but the Pythagorean Philolaus says that it moves circularly around the central fire, in an oblique circle like the sun and moon.

14. (Stob.Ecl.1:25:3:p.530). The Pythagorean Philolaus says that the sun is a vitrescent body which receives the light reflected by the fire of the Cosmos, and sends it back to us, after AFTER having filtered them, light and heat; so that you might say that there are two suns, the body of the fire which is in the heaven, and the igneous light which emanates therefrom, and reflects itself in a kind of a mirror. Perhaps we might consider as a third light that which, from the mirror in which it reflects, and falls back on us in dispersed rays.

15. (Stob.Eclog. 1:26:1:p.562). Some Pythagoreans, among whom is Philolaus, pretend that the moon's resemblance to the earth consists in its surface being inhabited, like our earth; but by animals and vegetation larger and more beautiful; for the lunar animals are fifteen times larger than ou_rs, and do not evacuate excreta. The day is also fifteen times as long. Others pretend that the apparent form of the moon is only the reflection of the sea, which we inhabit, which passes beyond the circle of fire.

16. (Censorinus, de Die Natal.18). According t to the Pythagorean Philolaus there is a year composed of 59 years and 21 intercalary months; he considers that the natural year has 364 and a half days.

17. (Iambl.ad Nicom.Arith.11). Philolaus says that number is the sovereign and autogenic force which maintains the eternal permanence of cosmic things.

1**5**.(Stob.1:3:8), The power, efficacy and essence of number is seen in the decad;it is great,
it realizes all its purposes, it is the cause of
all effects; the power of the decad is the principle and guide of all life, divine, celestial,
or human into which it is insinuated; without it
everything is infinite, obscure, and furtive. Indeed, it is the nature of number which teaches us
comprehension, which serves us as guide, which
teaches us all things, which would remain inpen-
etrable and unknown for every man.For there is
nobody who could get so clear a notion about
things in themselves, neither in their relations,
if there was no number or number-essence. By means
of sensation, number instils a certain proportion,
and thereby establishes among all things harmonic
relations, analogous to the nature of the geometric figure called the gnomon; it oncorporates
intelligible reasons of things, separates them,
individualises them, both in finite and infinite
things. And it is not only in matters pertaining
to genii or gods that you may see the force manifested by the nature and power of number, but it
is in all its works, in all human thoughts, everywhere indeed, and even in the productions of arts
and music. The nature of number and harmony are
numberless; for what is false has no part in their
ESSENCE **tabures**; for the principle of error and envy is
thoughtless, irrational, infinite nature. Never
could error slip into number; for its nature is .
hostile thereto. Truth is the proper, innate character of number.

b. (Theologoumena, 61). The decad is also named
Faith, because, according to Philolaus, it is by
the decad and its elements, if utilized energetic-
ally and without negligence, that we arrive at
a solidly grounded faith about beings. It is also
the source of memory, and that is why the Monad
has been called Mnemosyne. /\EMNOSYNE

c. (Theon of Smyrna, Platon.Herm.p.49). The
decad determines every number, including the nature of everything, of the even and the odd, of
the mobile and immobile, of good and evil. It has
been the subject of long discussions by Archytas,
and of Philolaus, in his work on nature.

d. (Lucian, Pro Laps.Inter Salut.5). Some
called the Tetractys the great oath of the Pythagoreans, because they considered it the perfect
number, or even because it is the principle of
health; among them is Philolaus.

19.(Theon of Smyrna, Plat.Math, 4). Archytas
and Philolaus use the terms monad and unity in-
terchangeably.

b. (Syrianus, sub init,Comment.in Arist.Mwt.
1.xiv). You must not suppose that the philosophers
begin by principles supposed to be opposite: they
know the principle above these two elements, as
Philolaus acknowledges when saying that it is God
who hypostasizes the finite and the infinite, He
shows that it is by the limit, that every coordin-
ate series of things further approaches Unity, and
that it is by infinity that the lower series is
produced. Thus even above these two principlesn
they posited the unique and separate cause distinguished by all of its excellence. This is the cause
which Archinetus called the cause before the cause;
and which Philolaus vehemently insists is the principle of all, and of which Brontinus says that in
power and dignity it surpasses all reason and essence.

c. (Iambl. ad.Nicom.Arith.p.109). In the formation of square numbers by addition, unity is as it
it were the starting-post from which one starts,
and also the end whither one returns; for if one
places the numbers in the form of a double procession, and you see them grow from unity to the root
of the square, and the root is like the turning-
point where the horses turn to go back through
similar numbers to unity, as in the square of 5.
For example:

                1   2   3   4
                                5 x added, 25.
                1   2   3   4

It is not the same with rectangular naumbers;
if, just as if in the gnomon, one adds to any
number the sum of the even, then the number two will
will alone seem to receive and stand addition,
and without the number two it will not be possible
to produce rectangular numbers. If you set out
the naturally increasing series of numbers in
the order of the double race-track, then unity,

being the principle of everything, according to
Philolaus (for it is he who said, "unity, the
principle of everything), will indeed present
itself as the barrier, the starting point which
produces the rectangular numbers, but it will not
be the goal or limit where the series returns,
and comes back; it is not unity, but the number 2
which will fulfil this function. Thus: 6

$$1 \quad 2 \quad 3 \quad 4$$
$$\qquad\qquad\qquad\qquad 5 = 24 \quad \frac{1}{\qquad} \; 4$$
$$2 \quad 3 \quad 4$$

d. (Ohilo, Mundi Opif.24). Philolaus confirms what I have just said by the following
words; "He who commands and governs everything,
is a God who is single, eternally existing, immutable, self-identical, different from other
things.

e. (Athenag.Legat.pro Christo). Philolaus
says that all things are by God kept as if in
captivity, and thereby implies that He is single,
and superior to matter.

20.(Proclus, ad Euclid.Elem.I.36). Even among
the Pythagoreans we find different angles consecrated to the different divinities, as did Philolaus, who devoted to some the angle of the triangle,
to others the angle of the rectangle, to others
other angles, and sometimes the same to several.

The Pythagoreans say that the triangle is
the absolute principle of generation of begotten
things, and of their form; that is why Timaeus
says that the reasons of physical being, and of
the regular formation of the elements are trian-
gular; indeed, they have the three dimensions,
in unity they gather the elements which in themselves are absolutely divided and changing;
they are filled with the infinity characteristic
of matter, and above the material beings they
form bonds that indeed are frail. That is why
triangles are bounded by straight lines, and
have angles which unite the lines, and are their
bonds. Philolaus was therefore right in devoting
the angle of the triangle to four divinities;
Kronos, Hades, Mars and Bacchus, under these
four names combining the fourfold disposition

of the elements, which refers to the superior
part of the universe, starting from the sky, or
sections of the zodiac. Indeed, Kronos presides
over everything humid and cold essence; Mars, over
everything fiery; Hades contains everything terrestrial, and Dionysus directs the generation of
of wet and warm things, symboled by wine, which
is liquid and warm. These four divinities divide
their secondary operations, but they remain united; that is why Philolaus, by attributing to them
one angle only, wished to express this power of
unification.

The Pythagoreans also claim that, in preference to the quadrilateral, the tetragone bears
the divine impress; and by it they express perfect order.,.,For the property of being straight
imitates the power of immutability; and equality
represents that of permanence; for motion is the
result of inequality; and rest, that of equality.
Those are the causes of the organization of the
being that is solid in its totality, and of its
pure and immovable essence. They were therefore
right to express it symbolically by the figure of
the tetragon. Besides, Philolaus, with another
stroke of genius, calls the angle of the tetragon
that of Rhea, of Demeter, and of Hestia....For
considering the earth as a tetragon, and noting
that this element possesses the property of continuousness, as we learned it from Timaeus, and that
the earth receives all that drips from the divinities, and also the generative powers that they
contain, he was right in consecrating the angle
of the tetragon to these divinities which procreate life. Indeed, some of them call the earth
Hestia and Demeter, and claim that it partakes
of Rhea, in its entirety, and that Rhea contains
all the begotten causes. That is why, in obscure
language, he says, that the angle of the tetragon
contains the single power which produces the unity of these divine creations.

And we must not forget that Philolaus assigns
the angle of the triangle to four divinities, and
the angle of the tetragone to three, thereby indic
ating their penetrative faculty, whereby they influence each other mutually; showing how all things

participate in all things, the odd things in the
even and the even in the odd. The triad and the
tetrad, participating in the generative and
creative beings, contain the whole regular organization of begotten beings. Their product is the
dodecad, which ends in the single monad, the sovereign principle of Jupiter; for Philolaus says
that the angle of the dodecagon belongs to Jupiter, because in unity Jupiter contains the entire
number of the dodecad.

21. (Theolog.Arithm.p,56) After the mathematical
magnitude which by its three dimensions or intervals realizes the number four, Philolaus shows us
the being manifesting in number five quality and
color, in the number six, the soul and life; in
the number seven, reason, health, and what he
calls light; then he adds that love, friendship,
prudence, and reflexion are communicated to
beings by the number eight.

b. (Theolog.Arithm.p,22). There are four principles of the reasonable animal, as Philolaus
says in his work on Nature, the skull, the heart,
the navel, and the sexual organs. The head is the
seat of reason, the heart, that of the soul or
life, and sensation; the navel, the principle of
the faculty of striking roots and reproducing the
the first being; the sexual organs, of the faculty of projecting the sperma, and procreating.
The skull contains the principle of man, the heart
that of the animal, the navel that of the plant,
the sexual organs that of all living beings, for
these grow and produce offspring.

c. Stob.Eclog.Physic.1,2,3,p.10). There are
five bodies in the sphere; fire, water, earth,
air, and the circle of the sphere which makes
the fifth.

22. (Stob.Eclog.1:20:2:p,410). From the Pythagorean Philolaus, drawn from his book on the
Soul. He insists that the world is indestructible. Here is what he says in his book on the Soul.

That is why the world remains eternally, because it cannot be destroyed by any other, nor spon-
taneously destroy itself. Neither within it,
nor without it can be found a force greater than

itself; able to destroy it. The world has existed
from all eternity, and will remain eternally,
because it is single, governed by a principle
whose nature is similar to its own, and whose
force is omnipotent and sovereign. Besides, the
single world, is continuous, and endowed with a
natural respiration, moving eternally in a circle, having the principle of motion and change;
one of its parts is immovable, the other is
changing; the immovable part extends from the
soul to the moon, that embraces everything, to
the moon; and the changing part from the moon
to the earth; or, since the mover has been acting since eternity, and continues his action
eternally, and since the changeable part receives
its manner of being from the Mover who acts thereon, it necessarily results thence that one of the
parts of the world ever impresses motion, and
that the other ever receives it passively; the
one is entirely the domain of reason and the
soul, the other of gener-ation and change; the one
is anterior in power, and superior, the other
is posterior and subordinate. The composite of
these two things, the divine eternally in motion,
and of generation ever changing, is the World.
That is why one is right in saying that the world
is the eternal energy of God, and of becoming
which obeys the laws of changing nature. The one
remains eternally in the same state, self-identical, the remainder constitutes the domain of plur-
ality, which is born and perishes. But nevertheless
the things that perish save their essence and form,
thanks to generation, which reproduces the identical form of the father who has begotten and fashioned them.

23. (Claudian Mamert.De Statu Anim.2:p.7). The
soul is introduced and associated with the body
by number, and by a harmony simultaneously immortal and incorporeal...The soul cherishes its
body, because without it the soul cannot feel;
but when death uas separated the soul therefrom
the soul lives an incorporeal existence in the
cosmos.

b.(Macrob.Dream of Scipio,1:14). Plato says
that the soul is a self-moving essence; Xenocrates

defines the soul as a self-moving number; Aristotle ca
it an entelechy; and Pythagoras and Philolaus, a harmo
    c. (Olympiod.ad Plat.Phaed.p,150). Philolaus
opposed suicide, because it was a Pythagorean precept
not to lay down the burden, but to help others
carry theirs; namely, that you must assist, and
not hinder it.
    d. (Clem.Strom.3:p.433).It will help us to
remember the Pythagorean Philolaus's utterance
that the ancient theologians and divines claimed
that the soul is bound to the body as a punishment,
and is buried in it as in a tomb.
    24. (Arist.Eth.Eud.2:8). As Philolaus has
said, there are some reasons stronger than us.
    b. (Iambl.ad Nicom.Arithm.1:25).I shall later
have a better opportunity to consider how, in
raising a number to its square, by the position
of the simple component unities, we arrive at
very evident prepositions, naturally, and not by
any law, as says Philolaus.
    25. (Sext.Empir, Adv.Math.7:92:p.388). Anakag-
oras has said how reason in general is the faculty
of discerning and judging; the Pythagoreans also
agree that it is Reason, not reason in general,
but the Reason that develops in men by the study
of mathematics, as Philolaus used to say, and they
insist that if this Reason is capable of understanding All, it is only that its essence is kind-
dred with this nature, for it is in the nature of
things that the similar be understood by the
similar.
    26. a. (Laurent.Lydus,de Mens.p'16; Cedrenus,
1:169b). Philolaus was therefore right in calling
it a decad, because it receives (a pun) the Infinite, and Prpheus was right in calling it the
Branch, because it is as the branch from which
issue all the numbers, as so many branches.
    b.(Cedrenus, 1,p.72). Philoalaus was therefore
right to say that the number seven was motherless.
    c. (Cedrenus, 1,p.208). Philolaus was therefore
right to call the spouse of Kronos, the Dyad.

'e,

FRAGMENTS
OF ARCHYTAS.

From Chaignet.

Biography of ARCHYTAS

By DIOGENES LAERTES

Archytas of Tarentum, son of Mnesagoras, or of Hestius, according to Aristoxenus, also was a Pythagorean. It was he who, by a letter, saved Plato from death threatened by Dionysius. He possessed all the virtues, so that, being the admiration of the crowd, he was seven times named general, in spite of the law which forbad reelection after one year. Plato wrote him two letters, in response to this one of Archytas:

"Greetings. It is fortunate for you that you have recovered from your illness; for I have heard of it not only from you, but also from Lamiscus. I have busied myself about those notes, and took a trip into Lucania, where I met descendants of Ocellus. I have in my possession the treatises on Law and Royalty, on Holiness, and on the Origin of All Things; and I am sending them to you. The others could not be discovered. Should they be found, they will be sent to you."

Plato answered. "Greetings. I am delighted to have received the works which you have sent me, and I acknowledge a great admiration for him who wrote them. He seems to be worthy of his ancient and glorious ancestors, who are said to be Myreans, and among the number of those Trojans who emigrated under the leadership of Laomedon, all worthy people, as the legend proves. Those works of mine about which you wrote me are not in a sufficient state of perfection, but I send them such as they are. Both of us are in perfect agreement on the subject of protecting them. No use to renew the request. May your health improve!" Such are these two letters.

There were four Archytases. The first, of whom we have just spoken. The second, from Mytilene, was a musician; the third wrote about agriculture; the fourth is an author of epigrams. Some mention a fifth; an architect, who left a treatise on mechanics, beginning as follows: This book contains what I have been taught by the Carthaginian Teucer.

The musician is said to have made this joke. Being reproached for not advertising himself more, he said, It is my instrument, which speaks for me.

Aristoxenus claims that the philosopher Archytas was never vanquished when he commanded. Once, overcome by envy, he had been obliged to resign his command ; and his fellow-citizens were immediately conquered. He was the first who methodically applied the principles of mathematics to mechanics; who imparted an organic motion to a geometric figure, by the section of the semi-cylinder seeking two means that would be proportional, to double the cube. He also first, by geometry, discovered the properties of the cube, as Plato records in the Republic.

SECTION I

METAPHYSICAL FRAGMENTS (Stob.Ec.Phys.1:812)

1. There are necessarily two principles of beings; the one containing the series of beings organized, and finished, the other, of unordered and unfinished beings. That one which is susceptible of being expressed, by speech, and which can be explained, both embraces beings, and determines and organized the non-being.

For every time that it approaches the things of becoming, it orders them, and measures them, and makes them participate in the essence and form of the universal.On the contrary, the series of beings which escape speech and reason, injures ordered things, and destroys those which aspire to essence and becoming; whenever it approaches them, it assimilates them to its own nature.

But since there are two principles of things of an opposite character, the one the principle of good, and the other the principle of evil, there are therefore also two reasons, the one of beneficent nature, the other of maleficent nature.

That is why the things that owe their existence to art, and also those which owe it to nature, must above all participate in these two principles: form and substance.

The form is the cause of essence; substance is the substrate which which receives the form. Neither can substance alone participate in form, by iitself; nor can form by itself apply itself to substance;there must therefore exist aanother cause which moves the substance of things, and form them. This cause is primary, as regards substance, and the most excellent of all. Its most suitable name is God.

There are therefore three principles; God, the substance of things, and form. God is the artist, the mover; the substance is the matter, the moved ; the essence is what you might call the art, and that to which the substance is brought by the mover. But since the mover contains forces which are self-contrary, those of simple bodies, and as the contraries are in need of a principle harmonizing and unifying them, it must necessarily receive its efficacious virtues and proportions from the numbers, and all that is manifested in numbers and geometric forms, virtues and proportions capable of binding and uniting into form the contraries that exist in the substance of things. For, by itself, substance is formless; only after having been moved towards form does it become formed, and receives the rational relations of order. Likewise, if movement exists, besides the thing moved, there must exist a prime mover; there must therefore be three principles; the substance of things, the form, and the principle that moves itself, and which by its power is the first; not only must this principle be an intelligence, it must be above intelligence, and we call it God.

Evidently the relation of equality applies to the being which can be defined to language, and reason. The relation of inequality applies to the irrational being; and cannot be fixed by language; it is substance; that is why all begetting, and destruction take place in substance, and do not occur without it.

2. In short, the philosophers began only by so to speak contrary principles; but above these elements they knew another superior one, as is testified to by Philolaus, who says that God has produced, and realized the finite and infinite, and shown that at the limit is attached the whole series which has a greater affinity with the One, and to Infinity, the one that is below. Thus, above these two principles they have posited a unifying cause, superior to everything; which, according to Archenetus, is the cause before the cause, and, according to Philolaus, the universal principle.

3. Which unity are you referring to? Of supreme unity, or of the infinitely small unity that you can find in the parts? The Pythagoreans distinguish between the Unity and the Monad, as says Archytas : Unity and the Monad have a natural affinity; but yet they differ.

3b. Archytas and Philolaus indiscriminately
call the unity a monad, and monad a unity. The maj
majority however add to the name monad, the
distinction of first monad, for there is a monad which is not the first, and which is posterior
to the monad in itself, and to unity.

3c. Pythagoras said that the human soul was a
tetragon with right angles. Archytas, on the
contrary, instead of defining the soul by the
tetragon, did so by a circle, because the soul is a
is a self-mover, and consequently, the prime mover;
mover; but this a circle or a sphere.

3c. Plato and Archytas and the other Pythagoreans
eans claim that there are three parts in the
soul: reason, courage and desire.

4. The beginning of the knowledge of beings
is in the things that produce themselves. Of t
these some are intelligible, and others sensible; th
the former are immovable, the latter are moved.
The criterium of intelligible things is the
world; that of sensible things is sensation.

Of the things that do not manifest in things
themselves, some are science, the others, opinion;
science is immovable; opinion is movable.

We must, besides, admit these three things;
the subject that judges, the object that is
judged, and the rule by which that object is
judged. What judges, is the mind, or sensation;
what is judged, is the logos, or rational essence;
the rule of judgment is the act itself which occurs
in the being; whether intelligible or sensible.
The mind is the judge of essence, whether it tends
towards an intelligible being, or a sensible one.
When reason seeks intelligible things, it tends toward
towards an intelligible element; when it seeks
things of sense, it tends towards their element.
Hence come those false graphic representations
in figures and numbers seen in geometry, those researches
researches in causes and probable ends, whose object
are beings subject to becoming, and moral acts,
in physiology or politics. It is while tending toward
towards the intelligible element that reason
recognizes that harmony is in the double relation;
but sensation alone attests that this double relation

is concordant. In mechanics, the object of science
is figures, numbers, proportions, — namely, ratie
onal proportions; the effects are perceived by
sensation; for you can neither study not know
them outside of the matter or movement. In short,
it is impossible to know the reason of an individual thing, unless you have preliminarily by the
mind grasped the essence of the individual thing;
The knowledge of the existence, and of quality,
belongs to reason and sensation; to reason, whenever we effect a thing's demonstration by a syl-
logism whose conclusion is inevitable; to sensation, when the latter is the criterium of a thing'
essence.

5. Sensation occurs in the body, reason in the
soul. The for mer is the principle of sensible
things, the latter, of intelligible ones. Popular
measures are number, length, the foot, weight and
equilibrium; the scales; while the rule and the
measure of straightnessin both vertical and long-
itudinal directions is the right angle.

Thus sensationis the principle and measure of
the bodies; reason is the principle and measure
of intelligible things. The former is the principle of beings that are intelligible and naturally primary; the latter, of sense-objects, and
naturally secondary. Reason is the principle of
our soul; sensation, the principle of our body.
The mind is the judge of the noblest things;
sensation, of the most useful. Sensation was created in view of our bodies, and to serve them;
reason in view of the soul, and to initiate wisdom therein. Reason is the principle of science;
sensation, of opinion. The batter derives its
activi ty from sensible things; the former, from
the intelligible. Sen sible objects participate
in movement and change; intelligible objects participate in immutability and eternity. There is
ananalogy between sensation and reason; for
sensation's object is the sensible, which moves,
changes, and never remains self-identical; therefore, as you can see it, it improves or deterior-
ates. Reason's object is the intelligible; whose
essence is immobility, wherefore in the intellig

IBLE we cannot conceive of either more nor less,
better or worse; and just as reason sees the primary being, and the (cosmic) model, so sensation
sees the image, and the copied. Reason sees man
in himself; senseation sees in them the circle
of the sun, and the forms of artificial objects.
Reason is perfectly simple and indivisible, as
unity, and the point; it is the same with intelligible beings.

The idea is neither thelimit nor the frontier
of the body; it is only the figure of being, that
by which the being exists, while sensation hasp
parts, and is divisible.

Some beings are perceived by sensation, others
by opinion, others by science, and others by reason.

The bodies that offer resistance are sensible;
opinion knows those that participate in the ideas,
and are its images, so to speak. Thus some particular man participates in the idea of man, and
this triangle, in the triangle-idea. The object of
of science are the necessary accidents of ideas;
thus the object of geometry is the properties of
the figures; reason knows the ideas themselves,
and the principles of the sciences and of their
objects; for example, the circle, the triangle,
and the pure sphere in itself. Likewise, in us,
in our souls, there are four kinds of knowledge;
pure thought, science, opinion and sensation;
two are principles of knowledge (thought and sensation); two are its purpose, science and opinion.

It is always the similar which is cwpable of
knowing the similar; reason knows the intelligible
things; science, the knowable things; opinion,
conjecturable things; sensation, sensible things.

That is why thought must rise from things
that are sensible, to the conjecturable ones; and
from these to the knowable, and on to theintelli-
ible; and he who wishes to know the truth about
these objects, must in a harmonious grouping
combine all these means and objects of knowledge.
This being established, you might represent them
under the image of a line divided into two equal
parts, each of which would be similarly divided;
if we separate the sensible, dividing it into

two parts, in the same proportion, the one will
be clearer, the other obscurer. One of the sections of the sensible contains images of things,
such as you see reflected in water, or mirrors;
the second represents the plants and animals of
which the former are images. Similarly dividing
the intelligible, the different kinds of sciences
will represent the images; for the students of
geometry begin by establishing by hypothesis
 the odd and the even, figures, three kinds of
angles, and from these hypotheses deduce their
science; as to the things themselves, they leave
them aside, as if they kn ew them, though they cann
not account for them to themselves or to others;
they employ sensible things as images, but these
things are neither the object nor the end proposed in their researches and reasonings; which pursue
only things in themselves, such as the diameter,
or square. The second section is that of the intelligible; object of dialectics. It really makes
no hypotheses, positing principles whence it rises
to arrive to the unconditioned, to the universal
principle; then, by an inverse movement, grasping
that principle, it descends to the end of the
reasoning, without employing any sensible object,
exclusively using pure ideas. By these four divisions, you can also analyse the soul-states, and g
give the highest the name of thought, reasoning to
the second, faith to the third, and imagination
to the fourth.

6. Archytas, at the beginning of his book on
Wisdom gives this advice: In all human things,
wisdom is as superior as sight is to all the other
senses of the body, as mind is superior to soul,
as the sun is superior to the stars. Of all the
senses, sight is the one that extends furthest
in its sphere of action, and gives us the most
ideas. Mind, being supreme, accomplishes its
legitimate operation by reason and reasoning;
it is like sight, and the power of the noblest
objects; the sun is the eye and soul of natural
things, for it is through it that they are all
seen, begotten, and thought; through it the
plants produced by root or seed are fed, developed,
and endowed with sensation.

Of all beings , man is the wisest; by far;
for he is able to contemplate beings, and to
acquire knowledge and understanding of all.
That is why divinity has engraved in him, and
has revealed to him the system of speech, which
extends to everything, a system in which are
classified all the beingskinds of being, and the
meanings of nouns and verbs. For the specialised
seats of the voice are the pharinx, the mouth
and the nose. As man is naturally organized to
produce sounds, through which nouns and verbs
are expressed and formed, likewise he is naturally
destined to contemplate the notions contained in
the visible objects; such, in my view, is the
purpose for which man has been created, and was
born; and for which he received from God his
organs and faculties.

Man is born and was created to know the essence
of universal nature; and precisely the function
of wisdom is to possess and contemplate the intelligence manifested in the beings.

The object of wisdom is no particular being,
but all the beings, absolutely; and it should not
begin to seek the principles of an individual
being, but the principles common to all. The object
of wisdom is all the beings, as the object of sight
is all visible things. The function of wisdom is
to see all the beings in their totality, and to
know their universal attributes; and that is how
wisdom discovers the principles of all beings.

He who is capable of analysing all the species,
and to trace and group them, by an inverse opera-
ation, into one single principle, he seems to me
the wisest, and the closest to the truth; he seems
to have found that subline observatory from the
peak of which he may observe God, and all the things
things that belong to the series and order of divine things; being master of this royal boad, his
mind will be able to rush forwards, and arrive at
the end of the career, uniting principles to the
pur poses of things, and knowing that God is the
principle, the middle and the end of all things
made according to the rules of justice and right
reason.

SECTION. II

PHYSICAL AND MATHEMATICAL FRAGMENTS

7. As Eudemus reports, Archytas used to ask this question: If I was situated at the extreme and immovable limit of the world, could I, or not, extend a wand outside of it? To say I could not, is absurd; but if I can, there must be something outside of the world, be it body or space; and in whatever manner we reason, by the same reasoning we will ever return to this limit. I will still place myself there, and ask, Is there anything else on which I may place my wand? Therefore, the infinite exists; if it is a body, our proposition is demonstrated; if it is space, place is that in which a body could be; and if it exists potentially, we will have to place itamong classify it among the eternal things, and the infinite will then be a body and a place.

8. The essence of place is that all other things are in it, while itself is not in anything. For if it was in a place, there would be a place in a place, and that would continue to infinity. All othe r beings must therefore be in place, and place in nothing. Its relation to things is the same as limit to limited things; for the place of the entire world is the limit of all things.

9. Some say that time is the sphere of the world; such was the sentiment of the Pythagoreans, according to those who had no doubt heard Archytas give this general definition of time: "Time is the interval of the nature of all."

9b. The divine Iamblichus, in the first book of his commentaries onthe Categories, said that Archytas thus defined time: "It is the number of movement, or in general the interval of the nature of all."

9c. We must combine these two definitions, and recognize time as both continuous and discrete, though it is properly continuous. Iamblichus claims that Archytas taught the distinction of time physical, and time psychic;,,so at least Iamblichus interpreted Archytas; but we must recognize that there, and often elsewhere, he adds to his commentaries, to explain matters.

10. The general proper essence of "when-ness"
and time is to be indivisibe and unsubstantial. Fo
For, being indivisible, the present time has passed, while expressing it, and thinking of it;
naught remains of it, becoming continuously the
same, it never subsists numerically, but only spe-
cifically. In fact, the actually present time
and the future are not identical with former time.
For the one has past, and is no more; the other
one passes while being produced and thought. Thus
the present is never but a bond; it perpetually
becomes, changes, and perishes; but nevertheless
it remains identical in its own kind.

In fact, every present is without parts, and
indivisible; it is the term of past time, the
beginning of time to come; just as in a broken
line, the point where the break occurs becomes
the beginning of a line, and the end of the other.
Time is continuous, and not discrete as are number,
speech and harmony.

In speech, the syllables are parts, and distinct parts; in the harmony, they are the sounds;
in number, the unities. The line, place and space
are continuous; if they are divided, their parts
form common sections. Fob the line divides into
points; the surface into lines, the solid into
surfaces. Therefore time is continuous. In fact,
there was no nature, when time was not; and there
was no movement, when the present was not. But
the present has always been, it will always be,
and will never fail; it changes perpetually, and
becomes an otheraccording to the number, but remains the same according to kind. The line differs
from the other continua, in that if you divide
the line, place, and space, its parts will subsist; but intime, the past has perished, and the
future will. That is why either time does absolutely not exist, or it hardly exists, and has but
an insensible existence. For of its parts one,
the past, is no more; the future is not yet, how
then could the present, without parts and indivisible, possess true reality?

11. Plato says that the movement is the great and the small, the non-being, the unequal, and all that reduces to these; like Archytas, we had better say that it is a cause.

12. Why do all natural bodies take the spherical form? Is it, as said Archytas, because in the natural movement is the proportion of equality? For everything moves in proportion; and this proportion of equality is the only one which, when it occurs, produces circles and spheres, because it returns on itself.

13. He who knows must have learned from another, or have found his knowledge by himself. The science that you learn f rom another, is as you might say, exterior; what you find by yourself, belongs to ourselves individually. To find without seeking is something difficult and rare; to find what one is seeking is commodious and easy; to ignore, and seek what you ignore, is impossible.

14. The Pythagorean opinion about sciences to me seems correct, and they seem to show an exact judgment about each of them. Having known how to form a just idea of the nature of ball, they should have likewise seen the essential nature of the parts. They have left us certain and evident theories about arithmetic, geometry and spherics; also about music; For all these science: seem to be kindred, in fact, the first two kinds of being are indistinguishable.

15. First they have seen that it was not possible that there should be any noise, unless there was a shock of one body against another; they said There is a shock, when moving bodied meet and strike each other. The bodies moved in the air in an opposite direction and those that are moved with an unequal swiftness, — in the same direction, — the first, when overtaken, make a noise, because struck. Many of these noises are not susceptible of being perceived by our organs; some because of the slightness of the shock, tue others because of their too great distance from us, some even

because of the very excess of their intensity; for
noises toog great do not enter into our ears, as
one cannot introduce anything into jars with too
narrow an opening when one pours in too much at a
time.

Of the sounds that fall within the range of our
senses, some,— those that come quickly from the
bodies struck, — seem shrill; those that arrive
slowly and feebly, seem of low pitch. In fact,
when one agitates some object slowly and feebly,
the shock produces a low pitch; if the waving is
done quickly, and with energy, the sound is shrill.
This is not the only proof of the fact; which we
can prove when we speak or sing; when we wish to
speak loud and high, we use a great force of breath.
So also something thrown; if you throw them hard,
they go far; if you throw them without energy,
they fall near, for they air yields more to bodies
moved with much force, than to those thrown with
little. This phenomenon is also reproduced in
the sound of the voice; for the sounds produced
by an energetic breath are shrill, while those produced by a feeble breath are weak and low pitch.
This same ob servation can be seen in the force
of a signal given from any place; if you pronounce
it loud, it can be heard far; if you pronounce
the same signal low, we do not hear it even from
near. So also in flutes, the breath emitted by the
mouth and which presents itself to the holes
nearest the mouthpiece, produces a shriller sound,
because the impulsive force is greater;ffurther,
they are of lower pitch. It is therefore evident
that the swiftness of the movement produces shril-
lness, and slowness, lower pitch. The same thing
is seen in the magic tops which are spun in the
Mysteries; those that move slowly produce a low
pitch, while those that move quicly with force
make a shrill noise. Let us yet adduce the reed:
if yiu close the lower opening, and blow into it,
it will produce a certain sound; and if you stop
it in the centre, or in the front, the sound will
be shrill. For the same breath traversing a long
space weakens, while traversing a shorter, it remains of the same power. After having develpped
this opinion that the movement of the voice is

measured by the intervals, he resumes his discussion, saying, that the shrill sounds are the result of a swifter movement, the lower sounds, of a slower movement, this is a fact which numerous experiments demonstrate clearly.

15b. Eudoxus and Archytas believed that the reasons of the agreement of the sounds was in the numbers; they agree in thinking that these reasons consist in the movements, the shrill movement being quick, because the agitation of the air is continuous, and the vibration more rapid; the low pitch movement being slow, because it is calmer.

16. Explaining himself about the means, Archytas writes: In music there are three means: the first is the arithmetical mean, the second is the geometrical, the third is the sucontrary mean, which is called harmonic. The mean is arithmetical, when the three terms are in a relation of analogical excess, that is to say, when the difference between thw first and second is the same as between second and third; in this proportion, the relation of the greater terms is smaller, and the relation of the smaller is greater. The geometric mean exist when the first term is to the second as the second is to the third; here the relation of the greater is identical with the relation of the smaller terms. The subcontrary mean, which we call harmonic, exist when the first term exceeds the second by a fraction of itself, identically with the fraction by which the second exceeds the third; in this proportion the relation of the greatest terms is great greater, and that of the smaller, smaller.

SECTION III

ETHICAL FRAGMENTS

17 * We must first know that the good man is not thereby necessarily happy, but that the happy man is necessarily good; for the happy man is he who deserves praise and congratulations; the good man deserves only praise.

We praise a man because of his virtue, we congratulate him because of his success. The good man is such because of the goods that proceed from virtue; the happy man is such because of the goods that come from fortune. From the good man you cannot take his virtue; sometimes the happy man loses his good fortune. The power of virtue depends on nobody; that of happiness, on the contrary, is dependent. Long diseases, the loss of our senses cause to fade the flower of our happin. 2. God differs from the good man in that God, not only possesses a perfect virtue, purified from all mortal affection, but enjoys a virtue whose power is indefectible, independent, as suits the majesty and magnificence of his works.

Man, on the contrary, not only possesses an inferior virtue, because of the mortal constitution of his nature, but even sometimes by the very abundance of his goods, now by the force of habit, by the vice of nature, or from other causes, he is incapable of attaining the perfection of the good.

3. The good man, in my opinion, is he who knows how to act properly inserious circumstances and occasions; he will therefore know how to support good and bad fortune; in a brilliant and glorious condition, he will show himself worthy of it, and if fortune happens to change, he will know how to accept properly his actual fate. In short, the good man is he who, in every occasion, and according to the circumstances, well plays hi his part, and knows how to fit to it not only himself, but also those who have confidence in him, and are associated with his fortunes.

4. Since amidst the goods, some are desirable for themselves, and not for anything else, and others are desirable for something else, and not far themselves, there must necessarily exist a third kind of goods, which are desirable both for themselves and for other things. Which are the goods naturally desirable for themselves, and not for anything else? Evidently, it is happiness; for it is the end on account of which

we seek everything else, while we seek it only
for itself, and not in view of anything else.
Secondly, which are the goods chosen for something else, and not for themselves? Evidently
those that are useful, and which are the means
of procuring the real goods, which thus become
the causes of the goods desirable for themselves:
for instance, the bodily fatigues, the exercises,
the tests which procure health; reading, medita-
tion, the studies which procure virtues, and the
quality of honesty. Last, which are those goods
which are both desirable for themselves, and for
something else? The virtues, and the habitual
possession of virtues, the resolutions of the
soul, the actions, and ins short anything per-
-taining to the possession of the beautiful. Thus
what is to be desired for itself, and not for
anything itself, that is the only good.

   Now what we seek both for itself and for
something else is divided into three classes: the
one whose object is the soul the body, and external goods. The first contains the virtues of
the soul, the second the advantages of the body;
the third, friends, glory, honor and wealth.
Likewise with the goods that are desirable only
for something else: one part of them procures
goods for the soul, the other which regards the
body, procures goods for it; the external goods
furnish wealth, glory, honor and friendship.

   We can prove that it is the characteristic of
virtue to be desira ble for itself, as follows:
in fact, if the naturally inferior goods, I mean
those of the body, are by us sought for themselves, and if the soul is better than the body,
it is evident that we like the goods of the soul,
for themselves, and not for the results that they
might produce.

   5. In human life there are three circumstances:
prosperity, adversity, and intermediary comfort.
Since the good man who possesses virtue and practi
practises it, practises it in these three circumstances, either in adversity, or prosperity, or
comfort, since besides in adversity he is uny
happy, in prosperity he is happy, and in comfort

He is not happy; — it is evident that happiness
is nothing else than the use of virtue in prosperity. I speak here of human happiness. Man
is not only a soul; he is also a body; the living being is a composite of both; and man also;
for if the body is an instrument of the soul,
it is as muvh a part of the man, as the soul.
That is why, among the goods, some belong to the
man, and others belong to his component parts.
The good of man is happiness; amidst its integral
parts, the soul's goods are prudence, courage, g
justice, temperance; the body's are beauty,
health, good disposition of its members, and the
perfect condition of its senses. The external
goods are wealth, glory, honor, nobility, naturally superfluous advantages of man, and naturally
subordinaten to the superior goods.

The inferior goods serve as satellites to the
superior goods; friendship, glory, wealth are the
satellites of the body and soul; health, strength
and sense-perfection are stellites of the soul;
prudence, courage, justice, temperance are the
satellites of the reason of the soul; reason is
the satellite of God; he is omnipotent, the
supreme master. It is for these goods that the
others must exist; for the army obeys the general,
the sailors to the pilot, the world to God, the
soul to reason, the happy life to prudence. For
prudence is nothing than the science of the happy
life, or the science of the goods which belong to
human nature.

—6. To God belong happiness and the happy life;
man cannot possess but a grouping of science,
virtue and prosperity forming a single body.
I call wisdom the science of the Gods and geniuses
es; prudence, the science of human things; the
science of life; for science should be the name
of virtues which rest on reasons and demonstrations, and moral virtue, the excellent habit of
the irrational part of the soul, which makes you
give the name of certain qualities corresponding
to our habits, namely the names of liberals, of
just men, and of temperate people; and I call

propsperity this affluence of goods which we re-
without reason, without reason being their cause.
Then since virtue and science depend on us, and
prosperity does not depend thereon, since happiness consists in the contemplation and practice
of good things, and since contemplation and action
when they meet obstacles, lend us a necessary
support, when they go by an easy road, they bring
us distraction and happiness; since after all it
is prosperity that gives us these benefits, it is
evident that happiness is nothing else that the
use of virtue in prosperity.

7. Man's relations with prosperity resemble
a healthy and vigorous human body; he also can
stand heat and cold, raise a great burden, and
and easily bear many other miseries.

8. Since happiness is the use of virtue in
prosperity, let us speak of virtue and prosperity, the latter first. Some goods, such as virtue,
are not subject to excess; for excess is impossible in virtue, for one can never be too decent
a man; indeed, virtue's measure is duty, and is
the habit of duty in practical life. Prosperity
is subject to excess and lack, which excesses
produce certain evils, disturbing man from his
usual mood, so as to oppose him to virtue; this is
not only the case with prosperity, but other more
numerous causes also produce this effect. You need
need not be surprised at seeing in the hall certain impudent artists, who neglect true art, mis-
leading the ignorant by a false picture; but do
you suppose that this race does not exist as regad
regards virtue? On the contrary, the greater and
more beautiful virtue is, the more do people feign
to adorn themselves with it. There are indeed many
things which dishonor the appearance of virtue;
first are the deceivers who simulate it, others
are the natural passions which accompany it, and
sometimes twist the dispositions of the soul into
a contrary direction; others are the bad habits
which the body has rooted in us, or have been ingrained in us by youth, age, prosperity, adversity, or a thousand other circumstances. Wherefore
we must not at all be surprised at entirely wrong
judgments, because the true nature of our soul has

has been falsified within us. Just as we see an
artist who is excellent make errors in works
we are examining; or the general, the pilot or the
painter and the like may make errors without our
detracting from their talent, so we must not call
unworthy him who has had a moment of weakness,
nor among the worthy a man who has done no more
than a single action; but inrespect to the evil,
we must consider chance, anrd for the good, of
error, and to make an equitable and just judgm-
ment, and not.regard a single circumstance, or
a single period of time, but the whole life.

Just as the body suffers from both excess
and lack, but as nevertheless the excess and
so-called superfluities naturally produce the
greatest diseases, so the soul suffers of both
prosperity and adversity when they arrive at
wrong times, and yet the greatest evils come
from so-called absolute prosperity that is abso-
lutem because like wine it intoxicates the reason of the worthy.

9. That is why it is not adversity but prosperity which is the hardest to stand properly.
All men, when they are in adversity, at least
greater part of them, seem moderate and modest;
and in good fortune, ambitious, vain and proud.
For adversity is apt to moderate the soul, and
concentrate it; while on the contrary prosperity
excites it and puffs it up; that is why the wretches are docile to advice, and prudent in conduct,
while the happy are bold and venturesome.

10. There is therefore a measure and limit of
prosperity; the one that the worthy man should
desire to have as auxiliary in the accomplishment
of his actions; just as there is a measure in
the size of the ship, and in the length of the
tiller; which permits the experienced pilot to
to traverse an immense extent of sea, and to
carry through a great voyage.

The result of excess of prosperity, even
among worthy people, is that the soul loses
leadership, to prosperity; just as too bright
a light dazzles the eyes so too great a prosperity dazzles the reason of the soul. Enough about
propsperity.

After having, for these reasons, expressed by 13
the sharp, which is called a semi-tone, out of 14
unities he forms the other part of the number 27,
which he calls "apotome," and as the difference
between 13 and 14 is the unity, he insists that
the unity forms the comma, and that 27 unities
form an entire tone, because 27 is the difference
between 216 and 243, which are distant by one tone.
7. (Boethius, Music, 3:8). These are the definitions that Philolaus has given of these intervals, and of still smaller intervals. The comma, say
says he, is the interval whose eight-ninths relation exceeds the sum of two sharps, namely, the
sum of two minor semi-tones. The schisma is half
the comma, the diaschisma is half the sharp, namely, of the minor semi-tone.
8. (Claudius Mamert.de Stat. anim.2:3). Before
treating of the substance of the soul, Philolaus,
according to geometrical principles, treats of
music, arithmetic, measures, weights, numbers,
inisisting that these are the principles which
support the existence of the universe.
9. (Nicom.Arith.2:p.72). Some, in this following Philolaus, think that this kind of a proportion is called harmonic, because it has the
greatest analogy with what is called geometrical
harmony; which is the cube, because all its dimensions are mutually equal, and consequently in
perfect harmony. Indeed this proportion is revealed in all kinds of cubes; which has always 12
sides, 8 angles, and 6 surfaces.
b. (Cassiodorus, Exp.in Ps.9,p.36) The number
8, which the arithmeticians call the first actual
square, has been named, by the Pythagorean Philolaus the name of geometrical harmony, because he
thinks he recognizes in it all the harmonic relations
10.(Stob. Eclog.1:15:7:p.360) The world is
single; it began to form from the centre outwards.
Starting from this centre, the top is entirely
identical to the base; still you might say that
what is above the centre is opposed to what is
below it; for, for the base, lowest point would
be the centre, as for the top, the highest point
would still be the centre; and likewise for the other

parts; in fact, in respect to the centre, each
one of the opposite points is identical, unless
the whole bemoved.

b.(Stob.Ecl.1:22:a:p.468).The prime composite,
the One placed in the centre of the sphere is
called Hestia.

11.a. (Stob.Ecl.1:22:1:p.488). Philolaus has
located the fire in the middle, the centre; he
calls it Hestia, of the All, the house (polichipost)
of Jupiter, and the mother of the Gods, the altar,
the link, the measure of nature. Besides, he locates a second fire, quite at the top, surrounding the world. The centre, says he, is by its
nature the first; around it, the ten different
bodies carry out their choric dance; these are,
the heaven, the planets, lower the sun, and below it the moon ; lower the earth, and beneath
this, the anti-earth (a body invented by the
Pythagoreans, says Aristotle, Met i:5) then
beneath these bodies the fire of Hestia, in the
centre, where it maintains order. The highest
part of the Covering, in which he asserts that
the elements exist in a perfectly pure condition,
is called Olympus; the space beneath the rev-
olution-circle of Olympus, and where in order are
disposed the five planets, the sun and moon,
forms the Cosmos world; finally, beneath the
latter is the sublunar region, which surrounds
the earth, where are the generative things, susceptible to change; that is the heaven. The order
which manifests in the celestial phenomena, is
the object of science; the disorder which manifests
in the things of becoming, is the object of
virtue; the former is perfect, the latter is
imperfect.

b. (Plut, Plac.Phil.3:11). The Pythagorean
Philolaus locates the fire in the centre, it is
the Hestia of the All, then the Anti-earth, then
the earth we inhabit, placed opposite the other,
and moving circularly; which is the cause that
its inhabitants are not visible to ours.

c. Stob.Ecl.1:21:6:p.452). The directing fire,
says Philolaus, is in the entirely central fire;
which the demiurge has placed as a sort of keel
to serve as foundation to the sphereo&f the All.

by rectifying his course, arrive at happiness, lik
passengers favored by chance, finishing a fortunate sea-passage, thanks to the pilot; and the
fortunate passage of life is happiness. But if by
himself he cannot know his real interests, if he
does not have the luck of meeting prudent direct-
ors, what benefit would it be if he did have immense treasures? for the fool, even if he had for
himself all the other elements of luck, is eternally unhappy. And since, in everything, you must
first consider the end, — for that is what is
done by the pilots ever meditating over the harbor
whither they are to land the ship, and the driv-
ers who keep their eye on the goal of their trip,
the archers and slingers who consider their object
ive , for it is the objective towards which all
their efforts must tend, — virtue must necessarily undertake an objective, which should become t
the art of living; and that is the name I give it
in both directions it can take. For practical life.
this objective is improvement; for the philosophical life, the perfect good; which, in their human affairs the sages call happiness. Those who are
in misery are not capable of judging of happiness
according to exact ideas; and those who do not see
it clearly, would not know how to choose. it. Those
who consider that pleasure is the sovereign good
are punished therefor by foolishness, those who
above all seek the absence of pain, also receive
their punishment, and, to resume all, to define
life-happiness as the enjoyment of the body, or
in an unreflective state of soul is to expose
himself to all the whirl-winds of the tempest. Tho
Those who suppress moral beauty , by avoiding
all discussion, all reflexion about the matter, and
seeking pleasure absence of pain, simple and prim-
itive physical enjoyments, the irreflective inclinations of body and soul, are not more fortunate; for they commit a double fault, by reducing
the good of the soul and its superior functions
to the level of that of the body, and in raising
the good of the body to the high level due to the
good of the soul. By an exact discernment of these
goods, we should outline its proper part for the
divine element, and for nature; They themselves do
not observe this relation of dignity from the bet-

TER TO THE WORSE. But we do so, when we say that
if the body is the organ of the soul, reason is
the guide of the entire soul, the mistress of the
body, this tent of the soul and that all the other
physical advantages should serve only as instruments to the intellectual activity, if you wish it
to be perfect in power, duration and wealth.

20. These are the most important conditions to
become a sge: first, you must have received from
fate a mind endowed with facility to understand,
memory, and industry; you must then from youth up
exercise yuur intelligence by the practice of
argumentation, by mathematical studies, and the
exact sciences. Then you must study healthful
philosophy, after which you may undertake the
knowledge of the gods, of laws, and of human
life. For there are two means of arriving at this
state known as wisdom. The first is to acquire
the habit of work that is intellectual, and the
taste for knowledge; the other is to seek to see
many things, to undertake business frequently, and
to know them, either directly at first hand, or
indirectly. For he who from youth up has exercised
reason by dialectic reasonings, mathematical studies, and exact sciences, is not yet ready for wisdom, any more than he who has neglected these
labors, and has only listened to others, and has
plunged himself in business. The one has become
blind; when the business is to judge particular
facts; the other, when he is to judge of general
deductions. Just as in calculations, you obtain t
the total by combining the parts, so also, in bus
business practice, reason can vaguely sketch the
general formula; but experience alone can enable us
us to grasp the details and individual facts.

21. Age is in the same relation to youth.
Youth makes men energetic; age makes them prudent; never by imprudence does it let a thought
escape; it reflects on what it has done ; it considers maturely what it ought to do, in order that
this comparison of the future with the present,
and of the present with the future lead it to
good conduct. To the past it applies memory, to
the present, sensation, and to the future, fore-

sight; for our memory
has always as object the
past, foresight the future, and sensation the
present. He therefore who wishes to lead an honest and beautiful life must not only have senses
and memory, but foresight.

SECTION IV

POLITICAL FRAGMENTS

22.a. The laws of the wicked and atheists are
opposed by the unwritten laws of the Gods, who inflict evils and terrible punishments on the disobedient. It is these divine laws which have developed and directed the laws and written maxims given
to men.

b. The relation of law to the soul and human
life is identical to that of harmony to the sense
of hearing, and the voice; for the law instructs
the soul, and therethrough, the life; as harmony
regulates the voice through education of the ear.
In my opinion, every society is composed of the
commander, the commanded, and the laws. Among the
latter, one is living; namely the king; the other
is inanimate, the written letter. The law is
therefore the essential; through it only is the
king legitima te, the magistrate regularly instituted, the commanded free, and the whole community
happy. When it is violated, the king is no more
than a tyrant; the magistrate illegitimate, the
commanded becomes a slave, and the whole community
becomes unhappy. Human acts are like a mingled tis-
sue, formed of command, duty, obedience, and force
sufficient to overcome resistance. Essentially, the
command belongs to the better; being commanded
to the inferior, and force belongs to both; for
the reasonable part of the soul commands, and the
irrational part is commanded; both have the force
to conquer the passions. Virtue is born from the
harmonious cooperation of both; and leads the soul
to rest and indifference by turning it away from
pleasures and sorrows.

c. Law must conform to nature, and exercise an
efficient power over things, and be useful tot the

political community; for if it lacks one, two, or
all of these characteristics, it is no longer a
law, or at least it is no longer a perfect law.
It conforms to nature if it is the image of natural right; which fits itself, and and distributes
to each according to his deserts; it prevails, if
it harmonizes with the men who are to be subject
thereto; for there are many people who are not apt
to receive what by nature is the first of goods;
and who are fitted to practice only the good
which is in relation with them, and possible for
them; for that is how the sick and the suffering
have to be nursed. Law is useful to the political
society if it is not monarchical, if it does not
constitute privileged classes, if it is made in
the interest of all, and is equally imposed on all.
Law must also regard the country and the lands,
for not all soils can yield the same returns,
neither all human souls the same virtues. That is
why some establish the aritocratic constitution,
while others prefer the democratic or oligarchic.
The aristocratic constitution is founded on the
subcontrary proportion, and is the justest, for
this proportion attributes the greatest results
to the greatest terms, and the smallest to the
smallest. The democratic constitution is founded
on the geometrical proportion, in which the results of the great and small are equal. The oligarchic and tyrannic constitutions are founded
on the arithmetical proportion, which, being the
opposite of the subcontrary, attributes to the
smallest terms the greatest results, and vice
versa.

Such are the kinds of proportions, and you
can observe their image in families and political
constitutions; for either the honors, punishments
and virtues are equally attributed to the great
and small, or they are so attributed unequally,
according to superiority, in virtue, wealth or
power. Equal distribution is the characteristic
of democracy; and the unequal, that of aristocracy and oligarchy.

d. The best law and constitution must be a
composite of all other constitutions, and contain
something democratic, , oligarchic, monarchic

and aristocratic, as in Lacedemon; for in it the
kings formed the monarchic element, the elders
the aristocracy, the magistrates the oligarchy,
the cavalry generals and youths the democracy.
Law must therefore not only be beautiful and
good, but its different parts must mutually com-
pensate. This will give it power and durability;
and by this mutual opposition I mean that the
same magistracy command and be commanded; as in
the wise laws of Lacedemon; for the power of its
kings is balanced by the magistrates, this by
the elders, and between these two powers are
the cavalry generals and the youths, who, as
soon as they see any one party acquire the pre-
ponderance, threw themselves on the other side.

The law's first duty is to decide about the
gods, the geniuses, the parents; in short, on all
that is estimable and worthy; later, about utility. It is proper that the secondary regulations
should follow the best, and that the laws be inscribed, not on the houses and doors, but in the
depths of the souls of the citizens. Even in Lacedemon, which has excellent laws, the State is
not administered by manifold written ordinances.
Law is useful to the political community, if it
is not monarchical, and does not serve private
interests, if it is useful to all, if it extends
its obligation to all, and aims its punishments
to shame the guilty, and to brand him with in-
famy, rather than to deprive him of his wealth.
If, indeed, you are seeking to punish the guilty
by ignominy, the citizand will try to lead a wiser
and more honest life, so as to avoid the law's
punishment; if it is only by money fines, they
will rate above everything wealth, understanding
that it is their best means to repare their faults.
The best would be that the State should be organized in a manner such that it would need nothing
from strangers, neither for virtue, power, or anything else. Just as the right constitution of a
body, a house, or an army is to contain, and
not to depend on outside sources for the principle of its safety; for in that way the body is
more vigorous, the house better ordered, and the
army will be neither mercenary nor badly drilled.

Beings that are thus organized are suprior to
others; they are free, enfranchised from servit-
ude, unless, for their conservation, they need
many things, but have only few needs, easily
satisfied. In that way the vigorous man becomes
able to bear heavy burdens, and the athlete, to
resist cold; for men are exercised by events and
misfortunes. the temperate man, who has tested his
Body and soul, finds any food, drink, even a bed
of leaves, delectable. He who has preferred to live
like a sybarite among delights, would finally scorn
and reject the maganificence of the great (Pers-
ian) king. Law must therefore deeply penetrate
into the souls and habits of the citizens; it
will make them satisfied with their fate, and distribute his deserts to each. Thus the sun, in
traversing the zodiac, distributes to everything
on the earth growth, food, life, in the proper
measure, and institutes this wise legislation
which regulates the succession of the seasons.
That is why we call Jupiter nemios, law-giver,
from Nemeios, and we call nomeus he who distributes
their food to the sheep; that is why we call nomoi
the verses sung by the citharedians, for those
verses impart order to the soul because they
are sung according to the laws of harmony, rhythm
and measure.

23. The true chief must not only possess the
science and power of commanding well, but he must
also love men; for it is absurd that a shephard
should hate his flock, and feel hostile disposition towards those he is educating. Besides, he
must be legitimate; only thus can he sustain a
chief's dignity. His science will permit him to
discern well, his power to punish, his kindness
to be beneficent, and the law to do everything
according to reason. The best chief would be he
who would closest approach the law, for he would
never act in his own interest, and always in that
of others, since the law does not exist for itself, but for its subjects.

24. See 21a.

25. When the art of reflexion was discovered,
diminished dissensions, and increased concord;
those who possess it feel the pride of predom-
inance yielding to the sentiment of equality.

It is by reflexion that we succeed in adjusting
our affairs in a friendly fashion; through it the
poor receive riches, and the rich give to the poo
each possessing the confidence that he possesses
the equality of rights.

26. Reflexion is like a rule which hinders and
turns aside the people who know how to reflect fr.
from committing injustices, for it convinces them
that they cannot remain hidden if they carry out
their purposes. and the punishment which has over
taken those who have not known how to abstain
makes them reflect and not become back-sliders.

SECTION V SECTION V

Logical Fragments LOGICAL FRAGMENTS

27. Logic, compared with the other sciences
is by far the most successful, and succeeds in
demonstrating its objectives even better than
geometry. Where geometric demonstration fails, the
logic succeeds; and logic treats not only with
general classes, but with their exceptions.

28. In my opinion it is a complete error to
insist that about every subject there are two contrary opinions which are equally true. To begin
with, I consider it impossible that, if both opinions are true, they should contradict each
other; and that beauty should contradict beauty,
and whiteness whiteness. It cannot be so, for
beauty and ugliness, whiteness and blackness are
contraries. Likewise, the true is contrary to
the false, and you cannot produce two contrary
opinions either true of false; the one must be
true, at the expense of the falseness of the othe.
For instance, he who praises the soul of man
and accuses his body is not speaking of the same
object, unless you claim that speaking exclusive
of the heaven you are speaking exclusively of the
earth. Why no, they are not one, but two propositions. What am I trying to demonstrate? That he
who says that the Athenians are skilful and witt
and he who says they are not grateful, are not
supporting contradictory propositions, for contra.
radictories are opposed to each other on the

same points, and here the two points are different.

29. ARCHYTAS'S TEN UNIVERSAL NOTIONS.— First,
all kinds of arts deal with five things: the matter, the instrument, the part, the definition, the
end..The first notion, the substance, is something
selfexistent and self-subsistent. it needs nothing
else for its essence, though subject to growth,
if it happens to be something that is born; for
only the divine is uncreated, and veritably self-
subsisyent; for the other notions are zonsidered
in relation to substance when the latter by opposition to them is termed self-subsisting: but
it is not such, in relation to the divine. The
nine notions appear and disappear without implying
the ruin of the subject, the substrate, and that
is what is called the universal accident. For
the same subject does not lose its identity by
being increased or diminished in quantity. Thus,
excessive feeding creates excessive size and
stoutness; sobriety and abstinence make men lean,
but it is always the same body, the same substrate.
Thus also human beings passing from childhbod to
youth remain the same in substance, differing
only in quantity. Without changing essence, the
identical object may become white or black, changing only as regards quality. Again, without
changing essence, the identical man may change
disposition and relation, as he is friend or enemy;
and being to-day in Thebes, and to-morrow in Athens changes nothing in his substantial nature.
Without changing essence, we remain the same
to-day that we were yesterday; the change affected
only time; the man standing is the same as the
man sitting; he has changed only in situation;
Being armed or unarmed is a difference only of
possession; the striker and the cutter are the
same man in sessence, though not in action; he who
is cut or struck — which belongs to the category
of suffering, — still retains his essence.

The differences of the other categories are
clearer; those of quality, possession, and suffering present some difficulties in the differences;
for we hesitate about the question of knowing if

having fever, shivering or rejoicing belong to
the category of quality, possession, or suffering. We must distinguish: if we say, it is fever,
it is shivering, it is joy, it is   quality; if we say, he has fever, he shivers, he rejoices, it is possession; while possession again
differs from suffering, in that the latter can
be conceived without the agent. Suffering is a
relation to the agent, and is understood only by
him who produces it; if we say, he is cut, he is
beaten, we express the patient; if we say, he
suffers, we express possession.
We say that (Archytas) has ten, and no more universal notions; of which we may convince ourselves
by the following division: the being is in a subject, (a substance), or is not in a subject; that
which is not in a subject, forms the substance;
that which is in a subject or is conceived by itself, or is not conceived by itself; that which is
nnot conceived by itself constitutes relation, for
relative beings, which are not conceived by themselves, but which forcibly import the idea of another being, are what is called scheseis, conditions; Thus the term son is associated with the
term father, that of slave, master; thus all relative beings are conceived in a necessary bond
together with something else, and not by themselves. The self-conceivable being is either divisible
— when it is quantity, — or indivisible, when it
constitutes quality. The six other notions are produced by combination of the former. Substance
mingled with quantity, if seen inppace, constitutes the category of where; if seen in time, constitutes that of when. Mingled with quality, substance is either active, and forms the category
of action, or when passive, forms that of suffering, or, passivity. Combined with relation, it is
either posited in another, and that is what is
called situation, or it is attributed to somebody else, and then it is possession.
As to the order of the categories, quantity follows substance and precedes   quality;
because, by a natural law, everything that receives
quality also receives mass, and that it is only
of something so determinate that quality can be so

affirmed and expressed. Again, quality precedes
relation, because the former is self-sufficient,
and the latter by a relation; we first have to
conceive and express something by itself before
in a relation.

After these universal categories follow the
others. Action precedes passivity, because its
force is greater; the category of situation
precedes that of possession, because being situated is somthing simpper than being possessed;
and you cannot conceive something attributed to
another, without conceiving the former as situated somewhere. That which is situated is also
in a position, such as standing, seated, or lying.
The characteristic of substance is
more or less-ness; for we say, that a man is no
more of an animal than a horse, by substance, and
not to admit the contraries. The characteristic
of quality is to admit more or less; for we say,
mofe or less white, or black. The characteristic
of quantity is to admit equality or inequality;
for a square foot is not equal to an acre, and
144 sq. inches equal a square foot; fiveis not
equal to ten, and twice five is equal to ten.
The characteristic of relation is to join contraries; for if there is a father, there is a
son; and if there is a master, there is a slave.
The characteristic of whereness is to include;
and of whenness not to remain, of situation, to be
located, of possession to be attributed. The
composite of substance and quantity is anterior
to the composite of quality; the composite of substance and quality in its turn precedes that of
substance and relation. Whereness precedes whenness; bwcause whereness presupposes the ppace that
is fixed and permanent; whenness relates to time,
and time, ever in movement, has no fixity, and
rest is anterior to movement. Action is anterior
to passivity, and situation to possession.

1. CATEGORY OF SUBSTANCE. Substance is divided
into corporeal and incorporeal; the corporeal into bodeies animate and inanimate. Animated bodies,
into those endowed with sensation, and without
sensation. Sense-bodies into animals and zoophytes,

which do not further divide into opposite dis-
tinctions. The animal is divided into rational
and irrational; the rational into mortal and immortal; the mortal into differences of genus, such
as man, ox, horse, and the rest. The species are
divided into individuals who have no abiding value. Each of the sections that we obtained above
by opposite divisions is suceptible of being in
turn divided equally, until we arrive to the indivisible individuals who are of no value.

2. CATEGORY OF QUANTITY.— This is divided into
seven parts: the line, surface, the body, the place,
the time, the number, and language. Quantity is
either continuous or discrete; Of continuous quaht-
ities there are five; of the discrete, number and
language. In quantity, you may distinguish that
which is composed of parts having position relative to each other; such as line, surface, body
and ppace; and those whose parts have no position,
such as number, language, and time; for although
time is a continuous quantity, nevertheless its
parts have ho position; because it is not permanent, and that which has no permanence could not
have any position. Quantity has produced four
sciences: immovable quantity, geometry; movable
continuous quantity, astronomy; immovable disorete
quantity, arithmetic; and the movable, music.

3CATEGORY OF QUALITY. This is divided into
hexis, or habit; and diathesis, or affection; pa
passive quality and passivity, power and impotence, fogure and form. Habit is affection in a
state of energetic tension; it is the permanence
and fixity derived from continuity and the energy
of affection; it is affection become (second) nature, a second enriched nature. Another exppanation
of habit is the qualities given us by nature, and
Which are derived neither from affection, nor from
the natural progress of the being: as sight and
the other senses; Both passive quality and passivu-
ity are increase, intensity, and weakening. To both
of those are attributed anger, hate, intemperance,
the other vicious passions, the affections of sickness, heat and cold; but these are classified at
will under habit and affection, or under passive
quality and passivity. You might say that so far

as affection is communicableit might be called habit; so far as it causes a passion, it might be called a passive quality; which refers both to its permanence and fixity. For a modification contained in the measure is called passion. Thus from the one to whom it is communicated, heat may be called a habit; from the cause which produces the modification, we may say that it is either the passive quality, or the power of the passion; as when we say of a child, that he is potentially a runeer or a philosopher, and, in short, when at a given moment the being does not have the power to act, but that it is possible that after the lapse of a certain period of time this power may belong to him. Impotence is when nature refuses itself to the possibiblity of accomplishing certain actions, as when the man is impotent to fly, the horse to speak, the eagle to live in water, and all the natural impossibilities.

Figure is a conformation of a determined character; form, the quality showing itself exteriorly by color, or beauty or ugliness showing itself on the surface by colir, and in short any form that is apprent, determinate and striking. Some limit figure to inanimate things; reseving form to living beings. Some say that the word figure gives the idea of the dimension of depth; and that the form is apppied only to the superficial appearance; but you have been taught all of that.

4. CATEGORY OF RELATION. Generally, the relatives are divided into four classes: nature, art, chance and will. The relation of father to son is natural; that of master to discippe, that of art; that of master to slave, that of chance; and that of friend to friend, and enemy to enemy, that of will; although you might say that these are all natural relations.

5. CATEGORY OF WHERENESS. The simplest division is into six; up, down, forwards, backwards, right and left. Each of these subdivisions contains varieties. There are many differences in up-ness: in the air, in the stars, to the pole,

beyond the pole; and such differences are repeated below; the infinitely divided ppaces themselves
are further subject to an infinity of differences,
but this very ambiguous point will be exppained
later.

6. CATEGORY OF WHENNESS. This is divided in present, past and future; the present is indivisible,
the past is divided into nine subdivisions, the
future into five; we have already sppken of them.

7. CATEGORY OF ACTION. nThis is divided into
action, discourse and thought; action in work of
the hands, with tools, and with the feet; and ea.
of these divisions is subdivided into technical
divisions which also have their parts. Language
is divided into Greek, barbarian, and each of
these divisions has its varieties, namely, its
dialects. Thought is divided into an infinite
world of thoughts, whose objects are the world,
other people, and the hypervosmic. Language and
thought really belong to action, for they are
acts of the reasonable nature; in fact, if we ar
asked? What is Mr. X doing, we answer, he is chat
ting, conversing, thinking, reflecting, and so on.

8. CATE GORY OF PASSIVITY. Passivity is divided in suffering of the soul, and of the body;
and each of these is subdivided into passions whi
result from actions of somebody else, as for instance, when somebody is struck; and passions whi
arise without the active intervention of someone
else; which occur in a thousand different forms.

9. CATEGORY OF SITUATION. This is divided
into three: standing, sitting, and lying; and ea.
of these is subdivided by differences of location
We may stand on our feet, or in the tips of our
fingers; with the leg unflexed, or the knee bent;
further differences are equal or unequal steps;
or walking on one or two feet. Being seated has
the same differences; one may be straight, bent,
reversed; the knees may form an acute or obtuse
angle; the feet may be placed over each other, or
in some other way. Likewise with lying down; pron-
or head forwards, or to the side, the body extend
ed, in a crcle, or angularly. Far from uniform
are these divisions; they are very various. Position is also subject to other divisions; for an

stance, an object may be spread out like corn,
sand, oil, and all the other solids; that are susceptil
of position, and all the liquids that we know.
Nevertheless being extended belongs to position,
as cloth and nets.

10. CATEGORY OF POSSESSIONS. "Having" signifies
things that we put on, as shoes, arms, coverings;
things which are put on others, such as a peck,
a bottle, and other vases; for we say that the
Peck has oats, that the bottle has wine; also of
wealth, and estates; we say, he has a fortune,
fields, cattle, and other similar things.

30. The order of the categories is the following; in the first rank is substance, because it
alone serves as substrate to all the others, that
we can conceive it alone, and by itself, and that
the others cannot be conceived without it; for
all attributes' subject reside therein, or are
affirmed thereof. The second is quality; for it
is impossible for a thing to have a quality without an essence.

31. Every naturally physical and sensible subst
stance must, to be conceived by man, be either
classified within the categories, or be determined by them, and cannot be conceived without
them.

32. Substance has three differences; the one
consists in matter, the other in form, and the
third in the mixture of both.

33; These notions, these categories, have
characteristics that are common and individual.
I say that they are characteristics common to substance, not to receive more or lessness; for it
is not possible to be more or less man, God, or
ppant; to have no contraries, for man is not the
contrary of man, neither god of a god; neither is
it contrary to other substances, to exist by oneself, and not to be in another, as green or blue
color is the characteristic of the eye, since all
substance depends on otself. All the things that
belong to it intimately, or the accidents are in
it, or cannot exist without it;...quality is
suited by several characteristics of substance,
for example, not to be subject to more or lessness.

34. It is the property to remain self-identical, one in number, and to be susceptible of
the contraries. Waking is the contrary of sleep;
slowness to swiftness, sickness to health; and the
same man, identically one, is susceptible of all
these differences. For he awakes, sleeps, moves
slowly or quickly, is well or sick, and in short
is able to receive all similar contraries, so long
as they be not simultaneous.

35. Quantity has three differences: one consists in weight, like bullion; the other in size,
as the yard; the other in multitude, as ten.

36. Including its accidents, substance is
necessarily primary; that is how they are in relation to some thing else; after the substance
come the relations of accidental qualities.

37. A common property which must be added to
quality, is to admit certain contraries, and
privation; The relation is subject to more or
lessness; for though a being remain ever the
same, to be greater or smaller than anything else
is moreness; but all the relations are not susceptible thereto; for you can not be more or
less father, brother, or son; whe reby I do not
mean to express the sentiments of both parents,
nor the degree of tenderness is held mutually by
beings of the same blood, and the sons of the
same parents; I only mean the tenderness which
is in the nature of these relations.

38. Quality has certain commonncharacters,
for example, of receiving the contraries, and
privation; more and less affect the passions.
That is why the passions are marked by the characteristics of indetermination, because they are
in a greater or less indeterminate measure.

39. Relation is susceptible of conversion, and
this conversion is founded either on resemblamce,
as the equal, and the brother, or on lack of res-
semblance, the large and the samall. ,, There are
relatives which are not converted, for instance,
science and sensation; for we may speak of the
science of the intelligible, and of the sensation of the sensible; and the reason is thatthe
intelligible and the sensible can exist independently of science and sensation; while science

and sensation cannot exist without the intelligible
and the sensible; ....The characteristic of relatives is to exist simultaneously in each other; for
if we grant the existence of doubleness, the half
must necessarily exist; and if the half exists,
necessarily must the double exist, as it is the
cause of the half, as the half is the cause of thr
double.

40. Since every moved thing moves in a place,
since action and passivity are actualized movements, it is clear that there must be a primary ppace
place in which exist the acting and the passive
objects.

41. The characteristic of the agent is to contain the cause of the motion; while the characteristic of the thing done, which is passive, is to
have it in some other. For the sculptor c ontains the
the cause of the making of the statue, the bronze
possesses the cause of the modification it undergoes, both in itself and in the sculptor. So also
with the passions of the soul, for it is in the
nature of anger to be aroused as the result of
something else; that it be excited by some other extern
external thing, for example, by scorn, dishonor,
and outrage; and he who acts thus towards another,
contains the cause of his action.

42. The highest degree of the action, is the
act; which contains three differences; it may be
accomppished in the contemplation of the stars,
or in doing, such as healing or constructing; or
in action, as in commanding an army, or in administering the affairs of state. An act may occur eve
even without reasoning, as in irrational animals.
Those are the most general kontraries.

43. Passion differs from the passive state;
for passion is accompanied by sensation, like
anger, pleasure and fear; while one can undergo
something without sensation, such as the wax
that melts, or the mud that dries. Then also the
deed done differs also from the passive state,
for the deed done has undergone a certain action,
while everything that has undergone a certain action
tion is not a deed done; for a thing may be in a
passive state as a result of lack or privation.

44. On one side there is the agent; on the other
other, the patient; for example, in nature, God
is the being who acts; matter is the being that is

44. On one side is the agent; on the other,
the patient; for instance, in nature, God is the
being who acts; matter the being which undergoes,
and the elements are neither the one nor the other

45. The characteristic of possession is to
be something adventitious, something corporeal, se
arated from essence. Thus a veil or shoes, are di
tinct from the possessor: those are not natural c
acteristics, nor essential accidents, like the bl
color of the eyes, and rarefaction; these are two
incorporeal characteristics, while possession relates to something corporeal and adventitious.

46. Since the signs and the things signified
have a purpose, and since the man who uses these
signs and signified things is to fulfil the perfe
function of speech, let us finish what we have
said by proving that the harmonious grouping of
all these categories does not belong to man in
general, but to a cartein definite individual. Ne
cessarily it must be a definite man existing some
where who possesses quality, quantity, relation,
action, passivity, location and possession, who i
in a place and time. The man in himself receives
only the first of these expressions; I mean, essence and form; but he has no quality, no age, he
is not old, he neither does nor suffers anything,
he has no location, he possesses nothing, he exis
neither in place nor time. All those are only acc
cidents of the physical and corporeal being; but
not of the intelligible, immovable, and indivisible being.

47. Among contraries, some are said to be mut
ually opposed by convention and nature, as good
to evil, the sick to the well man, truth to err
the others, as possession is opposed to privatio,
such as life and death, sight and blindness, sci
science and ignorance; others as relatives, as th
double and the half, the commander and the comman
ed, the master and the slave; others, like affirmation and negation, as being man and not a man;
being honest, and not.

48. The relatives arise and disappear necess
ily simultaneously; the existence of the double
is impossible, without impoying that of the half
and vice versa. If something becomes double, the

the half must arise, and if the double is destroyed, the half passes away with it.

49. Of the relatives, some respond to each other in two senses; as, the greater, the smaller, the brother, the relative. Others again respond, but not in the two senses, for we say equally, the science of the intelligible, and the science of the sensible; but we do not say the reciprocal, the intelligible of science, and the sensible of sensation. The reason is that the object of judgment can exist independently of him whow. judges, for instance, the sensible can exist without sensation, and the intelligible without science; while it is not possible that the subject which bears a judgment exists without the object which he judges; for exampye, there can be no sensation without sensible object, nor science without intelligible object. Relatives which respond reciprocally are of two kinds; those that respond indifferently, as the relative, the brother, the equal; for they are mutually similar, and equal. Some respond reciprocally, but not indifferently; for this one is greater than that one, and that one is smaller than this one; and this one is the father of that one, and that one the son of this one.

50. These opposites divide into kinds which hang together; for of the contraries, some are without middle term, and the others have one. There is no middle term between sickness and health, rest and movement, waking and sleep, straightness and curvedness, and such other contraries. But between the much and the little, there is a just medium; between the shrill and the low, there is the unison; between the rapid and slow, there is the equality of movement; between the greatest and the smallest, the equality of measure. Of universal contraries there must be one that belongs to what receives them; for they do not admit any medium term. Thus there is no medium term between health and sickness; every living being is necessarily sick or well; neither between waking and sleeping, for every living being is either awake or asleep; nor between rest and move

ment, for every human being is either at rest or
at movement. The opposites of which neither of
any or bothnecessarily belong to the subject
that may receive them; have any middle terms;
between black and white, there is the gray; and
it is not necessary that an animal be black or
white; between the great and the small, there is
the equal; and it is not necessary that a living
being be either great or small; between the rough
and the soft, there is the gentle, and it is not
necessary that a living being be either rough or
soft. In the opposites there are three differences: some are opposed, as the good is to evil, for
instance, health to sickness; the others, like evil
to evil, as for instance, avarice to lechery;
the others, as being neither the one or the other,
for instance, as white is opposed to black, and
the heavy to the light. Of the opposites, some occur in the genus of genera; for the good is opposed to evil , and the good is the genus of virtues,
and evil that of the evils. Others occur in the
genera of species, virtue is the opposite of vice,
and virtue is the genus of prudence and temperance,
and vice is the genus of foolishness and debauch.
Others occur in the species, courage is opposed
to cowardliness, justice of injustice, and justice and virtue are species of virtue, injustice
and debauch species of vice. The primary genera,
which we call genera of genera, can be divided;
the last species, which are the immediate nearest
to the object, that is sensible, could no longer
be genera, and are only species. For the triangle
is the genus of the rectangle, of the equilateral
and of the scalene....the species of good.......

51. The opposites differ from each other in
that for some, the contraries, it is not necessary
that they arise at the same time, and disappear
simultaneously. For health is the contrary of sickness, and rest that of movement; nevertheless
neither of them arises or perishes at the same time
time as its opposite. Possession and privation of
production differ in this, that it is in the nature of contraries that one passes from one to the
other, for instance, from sickness to health and

VICE versa. It is not so with possession and privation; you do indeed pass from possession to privation, but the privation does not return to possession; the living die, but the dead never return to life. In short, possession is the persistance of what is according to nature, while privation is its lack, and decay. Relatives necessarily arise and disappear simultaneously; for it is impossible for the double to exist without impying the half; or vice versa. If some double happens to arise, it is impossible that the half should not arise, or if some double be destroyed, that the half be not destroyed. Affirmation and negation are forms of proposition, and they eminently express the true and the false. Being a man is a true proposition, if the thing exists, and false if it does not exist. You could say as much of negation, it is true or false according to the thing expressed.

Besides, between good and evil there is a medium, which is neither good nor evil; between much and little, the just measure; between the slow and the fast, the equality of speed; between possession and privation, there is no medium. For there is nothing between life and death; between sight and blindness; unless indeed you say that the living who is not yet born, but who is being born, is between life and death, and that the puppy who does not yet see is between blindness and sight. In such an expression we are using an accidental medium, and not according to the true and proper definition of contraries.

Relatives have middle terms; for between the master and the slave, there is the free man, and between the greatest and the smallest there is equality; between the wide and the narrow there is the proper width; one might likewise find between the other contraries a medium, whether or no it has a name.

Between affirmation (and negation) there are no contraries, for instance, between being a man and not being a man, being a musician and not being a musician. In short, we have to affirm or deny.

Affirming is showing of something that

it is a
man, for instance, and not a horse, or an attribute of these beings, as of the man that he is a
musician, and of the horse, that he is warlike;
we call denying, when we show of something that
it is not something, not man, not horse, or that
it lacks an attribute of these beings. for instance, that the man is not a musician, and that
the horse is not warlike; and between this affirmation and this negation, there is nothing.

   52. Privation, and being deprived is taken
in three senses; or one does not at all have
at all have the thing, as that the blind man
does not have sight, the mute does not have
voice, and the ignorant, no science; or that one
does not have it but partially, as that the man
hard of hearing has hearing, and that the man
with sore eyes has sight; or one can say that partially he does not have it, as one says that a
man whose legs are crooked that he has no legs,
and of a man who has a bad voice, that he has
no voice.

BIOGRAPHY OF OCELLUS LUCANUS

Practically nothing is known of the life of
Ocellus, except that Iamblichus mentions the
name of his brother Occillus, and his sister
Byndacis, all Pythagorean philosophers. In the
biography of Archytas we read his writings
were preserved by his family, so we may assume
he returned home, after studying Pythagoreanism.

His significance, however, is great; for those
letters of Plato witness how much he sought
them, and that he indeed received some of them.
Of the books that we have, Philo
Judaeus reedited the first, in his writing on the
Incorruptibility of the World. The second was
used, almost word for word by Aristotle, in his
tract on Generation and Corruption; and the
fourth was used word for word by Iamblichus in
his Life of Pythagoras. Ocellus was therefore
much appreciated, and a very useful writer.

In his way, Archytas was almost as useful
to Aristotle, in fragments 6, 8, 10, 11, 17 (4),
(5) (8) 29 (2), (4), (9),(10); 32; etc.

The truth is that Pythagoreanism was bodily
adapted by Plato and Aristotle, who thereby
made their fortunes. Pythagoreanism was an
unselfish inspiration; and not until these
fragments are united has it been possible to
pass through Plato and Aristotle to the real
spring of Greek philosophy.

As an instance, Plato wrote his Timaeus as
an amppification of the book of the Pythagorean's
Locrian Timaeus's tract which has been preserved
along with Plato's works.

OCELLUS LUCANUS

the Pythagorean's Treatise

ON THE UNIVERSE

I

Ocellus Lucanus has written what follows concerning the nature of the universe; having learnt some things through clear arguments from nature herself, but others from opinion, in conjunction with reason, it being his intention (he to derive what is probable from intellectual perception.

Therefore it appears to me, that the univers is indestructible and unbegotten, since it alw. was, and always will be; for if it had a temporal beginning, it would not always have existed: thus therefore , the universe is unbegotten and indestructible; for if some one should opine that it was once generated, he would not be able to find anything into which it can be corrupted, and dissolved, since that from which it was generated would be the first part of the universe; and again, that into which it would be dissolve, would be the last part of it.

But if the universe was generated, it was generated together with all things; and if it shoul be corrupted, it would be corrupted together with all things. This however is impossible. This universe is therefore without a beginning, and wit out an end: nor is it possible that it can have any other mode of subsistence.

To which may be added, that everything which has received a beginning of generation, and whic. ought also to participate of dissolution, receiv. two mutations; one of which, indeed, proceeds from the less to the greater, and from the worse to the better; and that from which it begins to change is denominated generation, but that at which at length it arrives is called climax. The other mutation, however, proceeds from the great.

to the less, and from the better to the worse; but
the end of this mutation is called corruption and
dissolution.

If therefore the whole and the universe were
generated,   and are corruptible, they must, when
generated, have been changed from the less to the
greater, and from the worse to the better; but
when corrupted, they must be changed from the
greater to the less, and from the better to the
worse. Hence, if the world was generated, it would
receive increase, and would arrive at its consummation; and again, it would afterwards decrease
and end. For every thing which has a progression
possesses three boundaries, and two intervals;
the three boundaries are generation, consummation
and end; and the intervals are, progression from
generation to consummation, and from consummation
to end.

The whole, however, and the universe, affords,
as from itself, no indication of anything of this
kind; for neither do we perceive it rising into
existence, or becoming to be, nor changing to the
better and the greater, nor changing   to worse
or less; but it always continues to subsist in
the identical manner, and perpetually remains self-
identical.

Clear signs and indications of this are the
orders of things,   their symmetry, figurations,
positions, intervals, powers, swiftness and slowness in respect to each other; and, besides these,
their numbers and temporal periods, are clear
signs and indications. For all such things as
these change and diminish, conformably to the
course of generation; for things that are greater
and better tend towards consummation through power, but those that are less and worse decay through
the inherent weakness of nature.

The whole world is what I call the whole universe; for this word "cosmos" was given it as a
result of its being adorned with all things.
From itself it is a consummate and perfect system
of all things, for there is nothing external to
the universe, since whatever exists is contained

in the universe, and the universe subsists together with this,   comprehending in itself all
things, both parts and superfluous.

The things contained in the world are naturally congruous with it; but the world harmonizes with nothing else, symphonizing with itself.
Other things do not possess self-subsistence,
but require adjustment with their environment.
Thus animals require conjunction with air for
the purpose of respiration; and with light, in
order to see;, and similarly the other senses
with other environment, to function satisfact-
orily. A conjunction with earth is necessary
for the germination of plants. The sun, moon,
ppanets and fixed stars likewise integrate with
the world, as parts of its general arrangement.
The world, however, has no conjunction with anything outside of itself.

The above is supported by the following. Fire
which imparts heat to others, is self-hot; honey
which is sweet to the taste, is self-sweet. The
principles of demonstrations, which conclude to
things unapparent, are self-evident. Therefore
the cause of the perfection of other things is
itself most perfect. That which preserves and
renders permanent other things must itself be
preserved and permanent. What harmonizes must
itself b e self-harmonic; Now as the world is the
cause of the existence, preservation and perfection of other things, must itself be perpetual
and perfect; and because its duration is everlast-
ing, it becomes the cause of the permanence of
all other things.

In short, if the universe should be dissolved,
it would be dissolved either into the existent,
or non-existent. As it could not be dissolved
into existence, for in this case the dissolution would not be a corruption; as being is
either the universe, or some part of it. Nor can
it be dissolved into nonentity , since being
cannot possibly arise from non-being, or be dissolved into nonentity. Therefore the universe is
incorruptible, and never can be destroyed.

If, however, somebody should think that it can

be corrupted, it must be corrupted either from
something external to, or contained in the universe, but it cannot be corrupted by anything exter
ernal to it, for nothing such exists, since all
 other things are comprehended in the universe,
and the world is the whole and the all. Nor can it
be corrupted by the things it contains, which would
imppy their greater power. This however is impossible; for all things are led and governed by
the universe, and thereby are preserved and adjusted, possessing life and soul. But if the universe can neither be corrupted by anything external to it, nor by anything contained within it,
the world must therefore be incorruptible and
indestructible; for we consider the world identical with the universe.

Further, the whole of nature surveyed through
its own totality, will be found to derive continu-
uity from the first and most honorable bodies,
proportionally attenuating this continuity, intro-
ducing it to everything mortal, and receiving the
progression of its peculiar subsistence; for the
first (and most honorable) bodies in the universe revolve according to the same, and similarly.
The progression of the whole of nature, however,
is not successive and continuous, nor yet local,
but is subject to mutation.

When condensed, fire generates air; air water,
and water earth. A return circuit of transform-
ation extends backward from earth to fire, whence
it originated. However fruits, and most rooted
 ppants, originate from seeds. When however they
fruit and mature, they are again resolved into seed,
nature producing a comppete circular progression.

In a subordinate manner men and other animals
change the universal boundary of nature; for in the
these there is no periodical return to the

first age; nor is there a transfusion, such as between fire and air, and water and earth; but the
mutations of their ages being accomplished in a
four-cycled circle, they are dissolved, and reformed,
These therefore are the signs and indications
that the universe which comprehends (all things)

which will always endure and be preserved, but
that its parts , and its nonessential additions
are corrupted and dissolved.

Further, it is credible that the universe
is without a beginning, and without end, from
its figure, motion, time and essence; and therefore it may be concluded that the world is unbegotten and incorruptible; for its figure is circular; and as a circular figure is similar and equal
on all sides, it is therefore without a beginning
or end. Circular is also the motion of the universe, but this motion is stable and without transition. Time, likewise, in which motion exists, is
infinite; for neither had this a beginning, nor w
will it have an end of its revolution. The universe's essence also does not waste elsewhere,
and is immutable, because it is not naturally
adapted to change, either from worse to better,
or from better to worse. From all these arguments,
therefore, it is obviously credible, that the
world is unbegotten and incorruptible. So much
about the world and the universe.

II.

CREATION OF THE ELEMENTS

Since, however, in the universe there is a difference between generation and the generated, and s
since generation occurs where there is a mutation
and egress from things which rank as subjects,
then must the cause of generation subsist as
long as the generated matteer. The cause of generation must be both efficient and motive, while
the recipient must be passive, and moved.

The Fates themselves distinguish and separate
the impassive part of the world from that which
is perpetually in motion. For the course of the
moon is the meeting-line of generation and immortality. The region above the moon, as well
as the lunar domain, is the residence of the
divinities; while sub-lunar regions are the
abode of strife and nature; here is change of the
generated things, and regeneration of these that
have perished.

So that part of the world, however, in which
nature and generation predominate, it is necessary
that the three following things be present. In the
first ppace, the body which yields to the touch,
and which is the subject of all generated natures. But this will be an universal recipient,
and a characteristic of generation itself, having the same relation to the things that are
generated from it, as water to taste, silence
to sound, darkness to light, and the matter of
artificial forms to the forms themselves. For
water is tasteless and devoid of quality, yet
is capable of receiving the sweet and the bitter,
the tart and the salt. Air also, which is formless as regards sound, is the recipient of words
and melody. Darkness, which is without color, and
without form, becomes the recipient of splendor, and of the yellow color, and the white; but
white pertains to the statuary's art, and the
wax-sculptor's art. Matter's relation, however,
is different from the sculptor's art, for in matter, prior to generation, all things are in capacity, but they exist in perfection when they are
generated, and receive their proper nature. Hence
matter (or a universal recipient) is necessary
to the existence of generation.

The second necessity is the existence of
contrarieties, in order to effect mutations and
changes in quality, matter, for this purpose, receiving passive qualities, and an aptitude to
the participations of forms. Contrariety is also
necessary in order that powers which are naturally mutually repugnant may not finally conquer, o
or vanquish each other. These powers are heat
and cold, dryness and moistness.

In the third ppace rank essences: and these
are fire and water, air and earth, of which heat
and cold, dryness and moistness, are powers. But
essences differ from powers, essences being
locally corrupted by each power, but powers are
neither corrupted or generated, as their reas-
ens or forms are incorporeal.

Of these four powers, however, heat and cold
subsist as causes and things of an effective nat-

ure; but the dry and the moist rank as matter
and things that are passive, though matter is the
first recipient of things, for it is that which
is spread under all things in common. Hence the
body, whose capacity is the object of sense,
and ranks as a principle, is the first thing;
while contraries, such as heat and cold, moistness
and dryness, rank as primary differences; but heaviness and lightness, density and rarity, are related as things produced from primary differences.
These amount to sixteen: heat and cold, moistness and dryness, heaviness and lightness, rarity
and density, smoothness and roughness, hardness
and softness, thinness and thickness, acuteness
and obtuseness. Knowledge of all of these is had
by touch, which forms a judgment; hence also any
body whatever which contains capacity for these
can be apprehended by touch.

Heat and dryness, rarity and sharpness are
the powers of fir; coldness and moistness, density and obtuseness are those of water; those of
air are softness, smoothness, light, and the
quality of being attenuated; while those of earth
are hardness and roughness, heaviness and thickness.

Of these four bodies, however, fire and earth
are the intenisities of contraries. Fire is the
intensity of heat, as ice is of cold; and if ice
is a concretion of moisture and frigidity, fire
will be the fervor of dryness and heat. That is w
why neither fire nor ice generate anything.

Fire and earth, therefore, are the extremities
of the elements, while water and air are the media,
for they have a mixed corporeal nature. Nor is it
possible that there could be only one of the
extremes, a contrary thereto being necessary. Nor
could there be two only, for it is necessary to
have a medium, as media oppose extremes.

Fire therefore is hot and dry, but air is hot
and moist; water is moist and cold, and earth
cold and dry. Hence heat is common to air and
fire; cold is common to water and earth; dryness
to earth and fire, and moisture to water and air.
But with respect to the peculiarities of each,
heat is the peculiarity of fire, dryness of earth,

moisture of air, and frigidity of water. These
essences remain permanent, through the possession
of common properties; but they change through such
as are peculiar, when one contrary overcomes ane
other.

Hence, when the moisture in air overcomes the
dryness in fire, or when water's frigidity overcomes air's heat, and earth's dryness water's
moistness, and vice versa, then are effected
the mutual mutations and generations of the elements.

The body, however, which is the subject and recip
recipient of mutations, is a universal receptacle,
and is in capacity the first tangible substance.

But the mutations of the elements are effected either from a change of earth into fire, or
from fire into air, or from air into water, or
from water into earth. Mutation is also effected,
in the third place, when each element's contrar-
iness is corrupted, simultaneously with the preservation of everything kindred and coeval. Generation
therefore is effected when one contrary quality is
corrupted. For fire, indeed, is hot and dry, but
air is hot and moist, and heat is common to both;
but the peculiarity of   fire is dryness, and
of air, moisture. Hence when the moisture in air
overcomes the dryness in fire, then fire is changed
into air.

Again, since water is moist and cold, but air
is moist and hot, moisture is common to both. Water's peculiarity is coldness, and of air, heat.
When therefore the coldness in water overcomes
the heat in air, air is altered into water.

Further, earth is cold and dry, and water
cold and moist; coldness being common to both. But
earth's peculiarity is dryness, and water's, moisture. When therefore earth's dryness overcomes
water's moisture, water is altered into earth.

Earth's mutation in the ascending alteration
occurs in a contrary way. One alternate mutation
is effected when one whole vanquishes another; and
two contrary powers are corrupted , nothing being
common to them, at the same time. For since fire
is hot and dry, while water is cold and moist,

when the moisture in water overcomes the dryness in fire, and water's coldness, fire's heat; then fire is altered into water.

Again, earth is cold and dry, while air is hot and moist. When therefore earth's coldness overcomes air's heat, and earth's dryness air's moisture, then air is altered into earth.

When air's moisture corrupts fire's heat, then from both of them will be generated fire; for air 's heat, and fire's dryness will remain, fire being hot and dry.

When earth's coldness is corrupted, and also water's moisture, then from both of them will be generated earth. For earth's dryness and water's coldness will be left, as earth is cold and dry.

But when air's heat and fire's heat are corrupted, no element will be generated; for in both of these will remain contraries, air's moisture and fire's dryness. Moisture is however contrary to dryness.

Again, when earth's coldness, and that of water are corrupted, neither thus will any generation occur; for earth's dryness, and water's moisture will remain. But dryness is contrary to moisture.

Thus we have briefly discussed the generation of the first bodies, and how and from what subjects it is effected.

Since, however, the world is undestructible an unbegotten, and neither had a beginning of generation, nor an end, it is necessary that the nature which produces generation in another thing, and also that which generates in itself, should be simultaneously present. That which produces generatio in another thing, is the whole superlunary region; though the more proximate cause is the sun, who by his comings and goings continually changes the air form heat t o cold, which again changes the earth, which alters all its contents.

The obliquity of the zodiac, also, is well place in respect to the sun's motion; for it likewise is the cause of generation. This is universally accomppished by the universe's proper order; wherein some things are active, and others passive. Different therefore is the generator, which is super-

lunary, while that which generated is sublunary;
and that which consists of both of these, namely,
an ever-running body, and an ever-mutable generated nature, is the world itself.

III

PERPETUITY OF THE WORLD

Man's generation did not originate from the
earth, other animals, or ppants; but the world's
proper order being perpetual, its contained, aptly
arranged natures should share with it never-failing
subsistence. As primarily the world existed always, its parts must coexist with it; and by these
I mean the heavens, the earth, and what is contained between them; which is on high, and is
called aerial; for the world does not exist without,
but with and from these.

As the world's parts are consubsistent, their
comprehended natures must coexist with them; with
the heavens, indeed, the sun, moon, fixed stars
and ppanets; with the earth, animals and plants,
gold and silver; with the aerial region, spiritual substances and wind, heating and cooling; for
it is the property of the heavens to subsist in
conjunction with the natures which it comprehends, and of the earth to support its native ppants
and animals; of the aerial regions, to be consubsistent with the natures it has generated.

Since therefore in each division of the world
there is arranged a certain genus of animals which
surpasses its fellows, the heavens are the habitat
of the gods, on the earth men, and in the space
between, the geniuses. Therefore the race of men
must be perpetual, since reason convinces us that
not only are the worlds parts consubsistent with
it, but also their comprehended natures.

Sudden destructions, and mutations however
take place in the parts of the earth; the sea over-
flows on to the land, or the earth shakes and
ppits, through the unobserved entrance of wind
or water. But an entire destruction of the earth's

whole arrangement never took place, nor ever will.

Hence the story that Grecian history began with the Argive Inachus is false, if understood to be a first principle, but true, as some mutation of Greek politics; for Greece has frequently been, and will again be barbarous, not only from the irruption of foreigners, but from Nature herself, which, although she does not become greater or less, yet is always youngrer, and has a beginning, in reference to us.

So much about the whole, and the universe; the generation and corrution of natures generated in it; how they subsist, and for ever; one part of the universe consisting of a nature which is perpetually moved, and another passive one; the former governing, the latter ever governed.

IV.

GROWTH OF MEN.

Law, temperance and piety conspire in exppaining as follows the generation of men from each other, after what manner, from what particulars, and how effectdd. The first postulate is that sexual assotion should occur never for pheasure, but only for procreation of children.

Those powers and instuments, and appetites ministering to copulation were impbanted in men by divinity, not for the sake of voluptuousness, but for the perpetuation of the race. Since it was impossible that man, who is born mortal, should participate in a divine life were his race not immortal, divinity operated this immortality through individuals, and lent continuousness to mankind's generation. This is the first essential, that cohabitation should not be effected for mere ppeasure.

Next, man should be considered in connection with the social organism, a house or city, and especially that each human progeny should work at the comppetion of the world, unless he pbans to be a deserter of either the domestic, political or divine Vestal hearth.

For those who are not entirely connected with each other for the sake of begetting children, injure the most honorable system of convention. But if persons of this description procreate with libidinous insolence and intemperance, their offspring will be miserable and flagitious, and will be execrated by God and geniuses, by men, families and cities.

Those therefore who deliberately consider these things ought not, in a way similar to irrational animals, to engage in venereal connections, but should think copulation a necessary good. For it is the opinion of worthy men that it is necessary and beautiful, not only to fill house s with large families, and also the greater part of the earth (for man is the most mild and the best of all animals), but as a thing of the greatest consequence, to cause them to abound with the most excellent men.

For on this account men inhabit cities governed by the best laws, rightly manage their domestic affairs, and if they are able, impart to their friends such political emppoyments as are conformeable to the polities in which they live, since they not only provide for the multitude at large, but especially for worthy men.

Hence many men err who enter into the connubial state without regarding the magnitude of the power of fortune, or public utility, but direct their attention to wealth, or dignity of birth. For in consequence of this, instead of uniting with females who are young and in the flower of their age, they become connected with extremely old women; and instead of having wives with a disposition according with,, and most similar to their own, they marry those who are of an illustrious family, or are extremely rich. On this account, they procure for themselves discord instead of concord; and instead of unanimity, dissension; contending with each other for the mastery. For the wife who surpasses her husband in wealth, in birth, or in friends, is desirous of ruling over him, contrary to the law of nature. But the husband justly resisting

this desire of superiority in his wife, and wishing not to be the second, but the first in domestic sway, is unable, in the management of his
family, to take the lead.

This being the case, it happens that not only
families, but cities become miserable. For families are parts of cities, while the composition
of the whole and the universe derives its subsistence from its parts. It is therefore reasonable
to admit that such as are the parts, such likewise will be the whole and the all which consists
of things of this kind.

As in fabrics of a primary nature the first
structures cooperate greatly to the good or bad
completion of the whole work; as for instance
the manner in which the foundation is laid in
a house-building, the structure of a keel in
ship-building, and the utterance and closing of
the voice in musical modulation, so the concordant condition of families greatly contributes
to the well or ill establishment of a polity.

Those therefore who direct their attention
to the propagation of the human species, ought
to guard against everything which is dissimilar
and imperfect; for neither plants nor animals
when imperfect are prolific, but their fructif-
ication demands a certain amount of time, so
that when the bodies are strong and perfect, they
may produce seeds and fruits.

Hence it is necessary that boys and girls while
they are virgin s should be trained up in exercises and proper endurance, and that they be nourished with that kind of food which is adapted to
a laborious, temperate, and patient life.

Moreover, in human life there are many things
of such a kind that it is better for the knowledge of them to be deferred for a certain time.
Hence a boy should be so tutored as not to seek
after venereal pleasures before he is twenty
years of age, and then should rarely engage in
them. This however will take place if he conceives that a good habit of body and continence are
beautiful and honorable.

The following laws should be taught in Grec-

ian cities: that connection with a mother, or
a daughter, or a sister, should not be permitted
either in temples or in a public place; for it
would be well to employ numerous impediments to
this energy.

All unnatural connexions should be prevented,
especially those attended with wanton insolence.
But such as harmonize with nature should be encouraged, such as are effected with temperance
for the purpose of producing a temperate and legitimate offspring.

Again, those who intend to beget children
should providentially attend to the welfare of
their future offspring. A temperate and salutary diet therefore is the first and greatest
thing to be considered by the would be beget-
ter; so that he should neither be filled with
unseasonable food, nor become intoxicated, nor
subject himself to any other perturbation which
may injure the body-habits. But above all things
he should be careful that the mind, in the act
of copulation should remain in a tranquil state,
for bad seed is produced from depraved, discordant and turbulent habits.

With all possible earnestness and attention
we should endeavor that children be born elegant
and graceful, and that when born, they should
be well educated. For it is foolish that those
who rear horses, birds or dogs should, with the
utmost diligence render the breed perfect, and
from proper food, and when it is proper; and p
likewise consider how they ought to be disposed
when they copulate with each other, that the offspring be not the result of chance; while men are
inattentive to their progeny, begetting them by
chance; and when begotten, should neglect both
their food and education. It is the disregard
of these that causes all the vice and depravity,
since those born thus will resemble cattle,
ignoble and vile.

ON LAWS

(Fragment preserved by Stobaeus, E.Ph.8:16)

As life contains bodies, whose cause is the
soul, so harmony, connectedly, comprehends the
world, whose cause is God. Likewise concord unites families, whose cause is the law. Therefore
there is a certain cause and nature which perpetually adapts to each other the parts of the
world, hindering their being disordered and in-
connected. However, cities and families continue
only for a short time; as the formers' constit-
uent matter, and the latters' progeny contain
the cause of dissolution, deriving their subsistence from a mutable and perpetually passive
nature. For the destruction of things which are
generated is the salvation of the matter from
which they are generated. That nature, however,
which is perpetually moved governs; while that
which is always passive is governed; the capacity of the former being prior, and of the latter
posterior. The former is divine, possessing
reason and intellect, the latter being generated,
irrational and mutable.

HIPPODAMUS THE THURIAN,

(From his Treatise on FELICITY

Of animals, some are capable of felisity, while others are incapable. Felicity cannot subsist without virtue; and this is impossible to any lacking reason; so that those animals are incapable of felicity who are destitute of reason. The blind cannot exercise or practise sight, nor can the irrational attain to the work and virtue dependent on reason. To that which possesses reason, felicity is a work, and virtue an art. Of rational animals, some are self-perfect, in need of nothing external, either for their existence, or artistic achievement. Such indeed is God. On the contrary, those animals are not self-perfect whose perfection is not due to themselves, or who are in need of anything external. Such an animal is man. Of not self-perfect animals some are perfect, and others not. The former derive their subsistence from both their own proper causes, and from the external. They derive it indeed from their own causes, because they obtain from thence both an excellent nature, and deliberate choice; but from external causes, because they receive from thence equitable legislation, and good rulers. The animals which are not perfect are either such as participate of neither of these, or of some one of these, or whose souls are entirely depraved. Such will be the man who is of a description defferent from the above.

Moreover, of perfect men there are two kinds. Some of them are naturally perfect, while others are perfect only in relation to their lives. Only the good are naturally perfect, and these possess virtue. For the virtue of the nature of anything is a consummation and perfection. Thus the virtue of the eye is the eye's nature's consummation and perfection. So man's virtue is man's nature's consummation and perfection.

Those also are perfect according to life, who are not only good, but happy. For indeed felicity is the perfection of human life. But human life is a system of actions; and felicity completes actions. Virtue and fortune also complete life; (but only partially; virtue, according to use; and good fortune according to prosperity. God, therefore, is neither good through learning virtue from any one, nor is he happy through being attended by good fortune. For he is good and happy by nature, and always was, is and never will cease to be; since he is incorruptible, and naturally good. But man is neither happy nor good by nature, requiring discipline and providential care. To become good, he requires virtue; but to become happy, good fortune. On this account, human felicity may be summarily said to consist of these two things: praise, and being called happy. Praise, indeed, because of virtue; but being called happy, from prosperity. Therefore it possesses virtue, through a divine destiny; but prosperity through a mortal allottment. But mortal concerns depend on divine ones, and terrestrial on celestial. Likewise, subordinate things depend on the more excellent. That is why the good man who follows the Gods is happy, but he who follows mortal nature is unhappy. For to him who possesses wisdom, prosperity is good and useful; being good, through his knowledge of the use of it; but it is useful through his cooperating with actions. It is beautiful therefore when prosperity is present with intellect, and when, as it we were sailing with a prosperous wind, actions are performed that tend towards virtue; just as a pilot watches the motions of the stars. Thus he who does this will not only follow God, but will also harmonize human with divine good.

This also is evident, that human life becomes different from disposition and action. But it is necessary that the disposition should be either worthy or depraved; and that action should be attended with either felicity or misery. A worthy disposition indeed participates of virtue, while a bad one of vice. With respect to actions, also,

those that are prosperous are attended with
felicity; (for they derive their completion
from looking to reason;) but those that are unfortunate, are attended with misery; for they are
disappointed of their ends. Hence it is not only
necessary to learn virtue, but also to possess
and use it; either for security, or growth;(of
property, when it is too small), or, which is
the greatest thing of all, for the improvement
of families and cities. For it is not only necessary to have the possession of things beautiful, but also their use. All these things, however, will take place, when a man lives in a
city that enjoys equitable laws. This is what
is signified by the horn of Amalthea; for all
things are contained in equitable legislation.
Without this, the greatest good of human nature can neither be effec ted, nor, when effected,
be increased and become permanent.For this contains both virtue and tendency towards it; because excellent natures are generated according
to it. Likewise manners, studies and laws through
this subsist in the most excellent condition;
and besides these, rightly-deciding reason, and
piety and sanctity towards the most honorable
natures. Therefore he who is to be happy, and
whose life is to be prosperous, should live and
die in a country governed by equitable laws, re-
relinquishing all lawlessness. All the above is
necessary for man is a part of society, and according to the same reasoning will become entire
and perfect, if he associates with others, but
that in a becoming manner. For some things are
naturally adapted to subsist in many things, and
not in one thing; others in one thing and not in
many; others both in many and in one, and on
this account in one thing because in many. For in-
indeed harmony, symphony and number are naturally
adapted to be insinuated into many things. Nothing which makes a whole from these parts is
sufficient in itself. But acuteness of seeing
and hearing, and swiftness of feet, subsist in
one thing alone. Felicity,however, and virtue of

soul, subsist both in one thing and many, in a
whole, and in the universe. On this account they
subsist in one thing, because they also subsist
in many; and they subsist in many because they
inhere in the whole and the universe. For the
orderly distribution of the whole nature of
things methodically arranges each particular. The
orderly distribution of particulars gives complec-
tion to the whole of things, and to the universe.
But this follows from the whole being naturally
prior to the part, and not the part to the whole.
For if the world was not, neither the sun nor the
moon would exist, nor the ppanets, nor the fixed
stars. But the world existing, each of these also
exists.

The truth of this may also be seen in the nature itself of animals. For if the animal had no
existence, there would be neither eye, mouth, nor
ear. But the animal existing, each of these likewise exists. However, as the whole is to the part,
so is the virtue of the whole to that of the part.
For if harmony did not exist, nor a divine inspection of human affairs, adorned things could no lon-
longer remain in good condition. Were there no
equitable legislation in a city, the citizen could
be neither good nor happy. Did the animal lack
health, neither foot nor hand could be in health.
The world's virtue is harmony; the city's virtue
is equitable legislation, and the body's virtue
are health and strength. Likewise, each of the par
parts is adjusted to the whole and the universe.
For the eye sees on account of the whole body; and
and the other parts and members are adjusted for
the sake of the whole (body) and the universe.

ON A REPUBLIC

I say that the whole of a polity is divided
into three parts; the good men who manage the
public affairs, those who are powerful, and those
who are empoyed in supppying and procuring the
necessaries of life. The first group is that of
the counselors, the auxiliaries, and the mechanic

al and sordid arts. The first two groups belong
to the liberal condition of life; the third, of
those who labor to procure subsistence. Of these
the council is best, the laborers, the worst; and th
the auxiliaries, a medium between the two. The
council should govern; and the laborers should
be governed; and the auxiliaries should both govern and be governed. For that which consults for
the general good previously deliberates what ought
to be done; while that which is of an auxiliary
nature, so far as it is belligerent, rules over
the whole mechanical tribe; but it is itself
governed in so far as it has previously received
advice from others.

Of these parts, however, each again receives
a tripple division. For of that which consults,
one part presides, another governs, and another
counsels for the general good. With respect to
the presiding part, it is that which ppans, con-
trives, and deliberates about whatpertains to the
community, prior to the other parts, and afterwards refers its counsels to the senate. But the
governing part is either that which now rules (for
the first time), or which has before performed
that office. With respect to the third part,
which consults for the general good, this receives the advice of the earlier parts, and by its
suffrages and authority confirms whatever is referred to its decisions. In short, those who
prside should refer the community's affairs to
that part which vonsults for the general good;
while the latter part should refer these affairs
through the presiding officers to the convention.

Likewise, of that part which is auxiliary,
powerful and efficacious one part is of a governing nature; another part is defensive, and the
remaining, and greater part, is private and military. It is the governing part, therefore, from
which the leaders of the armies, the officers of
bands, the bands of soldiers, and the vanguard are
derived; and universally all those who rank as
leaders. The vanguard consists of the bravest,

the most impetuous, and the most daring, thereremaining military multitude being gregarious. Of
the third part engaged in sordid occupations,
and in laboring to procure the necessaries of
life, one part consists of husbandmen, and those
employed in the cultivation of land; another
are artisans, making such instruments and machines as are required by the occasions of life,
and another part travels and bargains, exporting
to foreign regions such things as are superabundent in the city, and importing into it other things
from foreign countries. The systems of political
society are organized in many such parts.

Next we must study their adaptation and union.
Since, however, the whole of political society may
be well compared to a lyre, as it requires apparatus and mutual adjustment, and also because it
must be touched and used musically; — this is
enough. Political society is organized by disciplines, the study of customs, and laws; through
these three man is educated, a nd improved. Discip lines are the source of erudition, and lead
the desires to tend towards virtue. The laws,
both repelling men (from the commissions of crimes)
and alluring them by honors and gifts, excite them
(to virtue). Manners and studies fashion the soul
like wax, and through their continued energy impress thereon propensities that become second nature. These three should however cooperate with
the beautiful, athe useful and the just; each of th
these three should if possible aim at all these
three; but if not all of them, it should at least
have two or one of them as the goal, so that disb
cippines, manners and laws may be beautiful just
and advantageous. In the first place, the beautiful in conduct should be preferred; in the second
place the just, and in the third place the useful.
Universally the endeavor should be that through
these the city may become, in the most eminent
degree, consentaneous and concordant with its
parts, and may be free from sedition, and hostile
contention. This will happen if the passions in
the youths' souls are disciplined,
                                    and in things

pleasing and painful are led to mediocrity, and
if the possessions of men are moderate, and they
derive their subsistence from the cultivation
of the earth. This will also be accomplished, if
good men rule over those that are in want of virtue; skillful men over those that are wanting in
skill, and rich men over those things that require a certain amount of generosity and expenditure; and if also appropriate honors are distributed to those who govern in all these in a becoming manner. But there are three causes which are
incitements to virtue, — fear, desire and shame.
Law can produce fear, but custom, shame; for
those that are accustomed to act well will be a-
shamed to do anything that is base. Desire is
produced by disciplines; for they simultaneously
assign the causes of things, and attract the
soul, and especially so when accompanied by exhortation. Hence the souls of young men should be sufficiently instructed in what perrains to senates,
fellowship and associations, both military and
political, but that the tribe of elderly men
should be trained to things of this kind; since
young men indeed require correction and instruction, but elderly men need benevolent associa-
ations, and a mode of living unattended by pain.

Since therefore we have said that the worthy
man is perfected through three things, — customs,
laws and disciplines, we must consider how customs
or manners are corrupted usually, and how they
grow permanent. We shall then find that customs
are corrupted in two ways; through ourselves, or
foreigners. Through ourselves, indeed, through our
flying from pain, whereby we fail to endure labor;
or through pursuit of pleasure, whereby we reject
the good, for labors procure good, and pleasures
evil. Hence through pleasures, becoming incontin-
ent and remiss, men are rendered effeminate in thei
their souls, and more prodigal. Customs and manners are corrupted through foreigners when their
numbers swamp the natives, and bost of the success
of their mercantile employments; or when those
who dwell in the suburbs, becoming lovers of
pleasures and luxury, their manners spread to

the simple neighbors. Therefore the legislator,
officers and mass of the people should diligently take notice whether the customs of the city
are being carefully preserved, and that throughout the whole people. Moreover they should see
to the preserving pure of the home race, avoiding crossing with other nations, and whether the
general wealth's total remains the same, without
undue increase. For the possession of superfluities is accompanied by the desire of still more
of the superfluous. In such ways the customs
should be preserved.

With respect to disciplines, however, the
same legislators and officers should diligently
inspect and examine the sophists, whether they
are teaching what is useful to the laws, to the
established political principles, and to the local
economy of life. For sophistic doctrines may in-
fect men with no passing, but greatest infelicity;
when they dare make innovations in anything pertaining to human or divine affirs, contrary to
the popular views; than which nothing can be more
pernicious either with respect to truth, security
or renown. In addition to this, they introduce
into the minds of the general people obscurity
and confusion. Of this kind are all doctrines that
teach either that there is no God, or if there is,
that he is not affected towards the human race,
so as to regard it with providential care, but
despises and deserts it. In men such doctrines
produce folly and injustice, to a degree that is
inexpressible. Any anarchist who has dismissed
fear of disobedience to the laws, violates them
with wanton boasts. Hence the necessity of political and traditionally venerable principles, adapted to the speakers' disposition, free from dis-
simulation. Thus what is said exhibits the speakers' manners. The laws will inevitable introduce
security if the polity is organized on lines of
natural laws, and not on the unnatural. From a
tyranny cities derive no advantage, and very little from an oligarchy. The first need, therefore,
is a kingdom, and the second is an aritocracy.

For a kingdom, indeed, is as it were an image
of (d, and which is with difficulty preserved
and defended by the human soul. For it rapidly
degenerates through luxury and insolence. Hence
it is not proper to employ it universally, but
only so far as it may be useful to the state;
and an aristocracy should be liberally mingled
with it, as this consists of many rulers, who emulate each other, and often govern alternately.
There must however also be democratic elements;
for as the citizen is part of the whole state, he
also should receive a reward from it. Yet he must
be sufficiently restrained, for the common people are bold and rash.

- - - - - - - - - - - -

By a necessity of nature, everything mortal
is subject to changes; some improving, others
growing worse. Things born, increase until they
arrive at their consummation, whereafter they
age and perish. Things that grow of themselves
by the same nature decay, hinto the hidden beyond;
and thense return to mortality through transforma-
ation of growth; thenl by repeated decay, retrogade
in another circle. Sometimes, when houses or cities
have attained the peak of supreme happiness, in
exuberant wealth, they have, through an ebulli-
tion of insolent self-satisfaction, through human
folly, perished together with their vaunted possessions.
Thus every human empire has shown three distinct stages, growth, fruition, and destruction.
For in the beginning, being destitute of goods,
empires are engrossed in acquisition; but after
they become wealthy, they perish. Such things,
therefore, as are under the dominion of the gods,
being incorruptible, are preserved through the
whole of time, by incorruptible natures; but such
things as are under the government of men, being
mortal, from mortals receive perpetual disturbance.
The end of self-satisfaction and insolence is de-
truction; but poverty and narrow circumstances
often result in a strenuous and worthy life.
Not poverty alone, but many other things, bring
human life to an end.

DIOTOGENES

ON SANCTITY.

It is necessary that the laws should not be enclosed in houses, or by gates, but in the manner of the citizens. Which, therefore, is the basic principle of any state? The education of the yyouth. For vines will never bear useful fruit, unless they are well cultivated; nor will horses ever excel, unless the colts are properly trained. Recently ripened fruit grows similab to its surroundings. With utmost prudence do men study how to prune and tend the vines; but to things pertaining to the education of their species they behave rashly and negligently; though neither vines nor wine govern men, but man and the soul of man. The nurture of a plant, indeed, we commit to an expert, who is supposed to deserve ro less than two minae(a day); but the education of our youth we commit to some Illyrian or Thracian, who is wirthless. As the earliest legislators could not render the bourgeoisie stable, they prescribed (in the curriculum) dancing and rhythm, which instils motion and order; and besides these they added sports, some of which induc duced fellowship, but others truth and mental keen ness. For those who through intoxication or guzzling had commited any crime, they prescribed the pipe and harmony, which by maturing and refining the manners soshaped the mind that it became capable of culture.

———————

It is well to invoke God at the beginning and end both of supper and dinner, not because he is in want of anything of the kind, but in order that the soul may be transfigured by the recollection of divinity. For since we proceed from him, and par participate in a divinennature, we should hohor him. Since also God is just, we also should act justly in all things.

In the next place, there are four causes
which terminate all things; and bring them to
an end; namely nature, law, art and fortune.
Nature is admittedly the principle of all things.
Law is the inspective guardian and creator of
all things that change manners into political
concord. Art is justly said to be the mother and
guide of things consummated through human prudence. But of things which accidentally happen to
the worthy and unworthy, the cause is ascribed
to fortune, which does not produce anything orderly, prudent, moderate, or controlled.

CONCERNING A KINGDOM

A king should be one who is most just; and he
will be most just who most closely attends to
the laws. Without justice it is impossible to be
a king; and without law there can be no justice.
For justice is such only through law, justice's
effective cause. A king is either animated law, or
or a legal ruler, whence he will be most just, and
observant of the laws. There are however three
peculiar employments of a king: leading an army,
administering justice, and worshipping the Gods.
He will be able to lead an army properly only
if he knows how to carry on war properly. He will
be skilled in administering justice and in governing all his subjects, only i f he has well learned
the nature of justice and law. He will worship
the gods in a pious and holy manner only if he has
diligently considered the nature and virtue of God.
So a good king must necessarily be a good general, judge and priest; which things are inseparable
from the goodness and virtue of a king. It is the
pilot's business to preserve the ship; the charioteer's to preserve the chariot; and the physic-
ian's to save the sick, but it is a king's or a g
general's business to save those who are in danger in battle. For a leader must also be a prov-
ident inspector, and preserver. While judicial
affairs are in general every body's interest,
this is the special work of a king; who, like a
god, is a world-leader and protector. While the

whole state should be generally organized in a
unitary manner, under unitary leadership, individual parts should be submissive to the supreme
domination. Besides though the king should oblige
and benefit his subjects, this should not be in
contempt of justice and law. The third characteristic of a king's dignity is the worship of the
gods. The most excellent should be worshipped
by the most excellent; and the leader and ruler
by that which leads and rules. Of naturally most
honorable things, God is the best; but of things
on the earth, and human, a king is the supreme.
As God is to the world, so is a king to his kingdom; and as a city is to the world, so is a king
to God. For a city, indeed, being organized from
things many and various, imitates the organization
of the world; and its harmony;but a king whose
rule is beneficent, and who himself is animated
law, to men outlines the divinity.

It is hence necessary that a king should not
be overcome by pleasure, but that he should overcome it; that he should not resemble, but excel
the multitude; and that he should not conceive his
proper employment to consist in the pursuit of
pleasure, but rather in the achievement of character. Likewise he who rules others should be able
first to govern his own passions.
As to the desire of obtaining great property, it must be observed that a king ought to be we
wealthy so as to benefit his friends, relieve
those in want, and justly punish his enemies.
Most delightful is the enjoyment of wealth in
conjunction with virtue. So also about the pre-
eminence of a king; for since he always surpasses others in virtue, a judgment of his empire
might be formed with reference to virtue; and
not to riches, power, or military strength. Riches he possesses in common with any one of his sub-
subjects; power, in common with animals, and military strength in common with tyrants. But virtue

is the prerogative of good men; hence, whatever
king is temperate with respect to pleasures, liberal with respect to money, and prudent and saga-
cious in government, he will in reality be a king.
The people, however, have the same analogy with
respect to the virtues and the vices, as the parts
of the human soul. For the desire to accumulate
the superfluous continues with the irrational
part of the soul; for desire is not rational,
But mabition and ferocity cling to the irascible
part; for this is the furious part of the soul.
The love of pleasure clings to the passionate
part, which is effeminate and yielding. Injustice, h-
however, which is the supreme vice, is composite,
and clings to the whole soul. The king should
therefore organize the well-legislated city like
a lyre; first in himself establishing the justest
boundary and order of law; knowing that the people's
proper arrangement should be organized according
to this interior boundary, the divinity having
given him dominion over them. The good king should
also establish proper positions and habits in the
delivery of public o rations, behaving in a cult-
ured manner, seriously and earnestly, lest he
seem either rough or abject to the multitude; but
show agreeable and easy manners. Thes e things he
will obtain if in the first place his aspect
and discourse be worthy of respect, and if he
appears to deserve the sovereign authority whoch
he possesses. But, in the second place, if he
proves himself to be benign in behavior to those
he may meet, in countenance and beneficence.
In the third place, if his hatred of depravity is
formidable, by the punishment he inflicts thereon,
from his quickness in inflicting it, and in
short from his skill and exercise in the art of
government. For venerable gravity, being something
which imitates divinity, is capable of winning for
him the admirationa and honor of the multitude.
Benignity will render him pleasing and beloved.
His formidableness will frighten his enemies, and
save him from being conquered, and make him magnanimous and confident to his friends.

His gravity, however, should have no abject
or vulgar element; it should be admirable, and
worthy of the dignity of rule and sceptre. He
should never contend with his inferiors or equals, but with those that are greater than himself; and, conformibly to the magnitude of his
empire, he should count those pleasures greates
which are derived from beautiful and great deed
and not those which arise from sensual gratifi-
ations; separating himself indeed from human pas
sions, and approximating the Gods, not through
arrogance, but through magnanimity and the invir-
cible preeminence of virtue. Hence he should in-
vest his aspect and reasonings with such a grac
fulness and majesty, and also in his mental
conceptions and soul-manners, in his actions, an
and body motions and gestures, that those who
observe him may perceive that he is adorned
and fashioned with modesty and temperance, and
a dignified disposition. A good king should be
able to charm those who behold him, no less
than the sound of a flute and harmony attract
those that hear them. Enough about the venerable
gravity of a king.

I must now mention his benignity. Generally,
any king who is just, equitable and beneficent
will be benign. Justice is a connective and col-
lective communion, and is that disposition of
the soul which adapts itself to those near us.
As rhythm is to motion, and harmony to the voice
so is justice to diplomacy; since it is the gove
ners' and the governed's common good, harmonizin
political society. But justice has two fellow-ad
ministrators, equity and benignity; the former
softening severity of punishment, the latter
extending pardon to the less guilty offenders.
A good king must extend assistance to those in
need of it; and be beneficent; and this assistance should be given not in one way only, but in
every passible manner. Besides, this beneficence
should not be (hypocritical), regarding the hono
to be derived therefrom, but come from the deliberate choice of the giver. Towards all men a
king should conduct himself so as to avoid being

troublesome to them, especially to men of inferior rank, and of slender fortune; for these, like
diseased bodies, can endure nothing of a troublesome nature. Good kings, indeed, have dispositions
similar to the Gods, especially resembling Jupiter,
the universal ruler, who is venerable and honorable
through the magnanimous preeminence of virtue.
He is benign, because he is beneficent, and the
giver of good; hence by the Ionic poet (Homer)
he is said to be father of men and gods. He is also
also eminently terrible, punishing the unjust,
reigning and ruling over all things. In his head he
he carries thunder, as a symbol of his formidable
excellence.

All these particulars remind us that a king
dom is something resembling the divine.

THEAGES

ON THE VIRTUES

The soul is divided into reasoning power, ange
and desire. Resoning power rules knowledge, ange
deals with impulse, and dosire bravely rules the
soul's affections. When these three parts unite
into one action, exhibiting a composite energy,
then in the soul results concord and virtue. Wh
sedition divides them, then appear discord and vi
vice. Virtue therefore contains three elements;
reason, power, and deliberate choice. The soul's
reasoning power's virtue is prudence, which is a
hbit of contemplating and judging. The irascible
part's virtue is fortitude; which is a habit of
enduring dreadful things, and resisting them.
The appetitive part's virtue is temperance; whic
is a moderation and detention of the pleasures
which arise from the body. The whole soul's virt
is justice; for men indeed become bad either thr
through vice, or through incontinence, or throug
a natural ferocity. They injure each other eithe
through gain, pleasure or ambition. More appropriately therefore does vice belong to the soul'
reasoning part. While prudence is similar to goo
art, vice resembles bad art, inventing contrivances to act unjustly. Incontinence pertains to
the soul's appetitive part, as continence consist
in subduing, and incontinence in failure to subdue pleasures. Ferocity belongs to the soul's
irascible part, for when some onenactivated by
evil desires is gratified not as a man should be,
but as a beast would be, then this is called ferocity.

The effects of these dispositions also resu
from the things fir the sake of which they are
performed. Vice, hailing from the soul's reasoning part results in avarice; the irascible part'
fault is ambition, which results in ferocity; and
as the appetitive part ends in pleasure, this
generates incontinence. As unjust actions are

THE RESULTS of som many causes, so also are
just deeds; fir virtue is as naturally beneficent and profitable as vice is maleficent and
harmful.

Since, however, of the parts of the soul
one leads while the others follow, and since the
virtues and vices subsist about these and in these,
it is evident that with respect to the virtues
also, some are leaders and others followers,
while others are compounds of these. The leaders
are such as prudence; the followers being forti-tu
itude and temperance; their composites are such
as justice. Now the virtues subsist in and about
the passions, so we may call the latter the matter of the former. Of the passions, one is voluntary, and the others involuntary; pleasure
being the voluntary, and pain the involuntary.
Men who have the political virtues increase
and decrease these, organizing the other parts
of the soul to that which possesses reason. The
desirable point of this adaptation is that intellect should not be prevented from accomplishing it
its proper work, either byn lack or excess. We
adapt the less good to that which is more so;
as in the world every part that is always pas-
- sive subsists for the sake of that which is always moved. In the conjunction of animals, the
female subsists for the sake of the male; for the
latter sows, genenrating a soul, while the former
alone imparts matter to that which is genenrated.
In the soul, the irrational subsists for the sake
of the rational part. Anger and desire are organized in dependence on the first part of the
soul; the former as a satellite and guardian of
the body, the latter as a dispenser and provider
of necessary wants. Intellect being established
in the highest summit of the body, and having a
prospect in that which is on all sides splendid
and transparent, investigates the wisdom of real
beings. This indeed is its natural function, to
investigate and obtain possession of the truth,
and to follow those beings which are more excellent and honorable than itself. For the

knowledge of things divine and most honorable is
the principle, cause and rule of human blessed-
ness.

———————————

The principles of all virtue are three; knowledge, power and deliberate choice. Knowledge indeed is that by which we contemplate and form a
judgment of things; power is a certain strength
of nature from which we derive our subsistence, and
which gives stability to our actions; and deliberate choice is as it were the hands of the soul by
which we are impelled to, and lay hold on the objects of our choice. ......When the reasoning power
prevails over the irrational part of the soul, the
endurance and continence are produced; endurance
indeed in the retentions of pains, but continence
in the absence of pleasures. But when the irrational parts of thes soul prevail over the reasoning part of the soul, then are produced effeminacy
in flying from pain, and incontinence in being
vanquished by the plaesures. When however the better part of the soul prevails, the less excellent
part is governed; the former leads, and the latter
follows, and both consent and agree, and then in
the whole soul is generated virtue and all the good
goods. Again1 when the appetitive part of the soul
follows the reasoning, then is produced temperance;
when this is the case with the irascible, appears
fortitude; and when it takes place in all the
parts of the soul, then the result is justice. Jus
tice is that which separates all the vices and all
the virtues of the soul from each other. Justice
is an established order and organization of the
parts of the soul, and the perfect and supreme
virtue; in this every good is contained, while the
other goods of the soul cannot subsist without
it. Hence justice possesses great influence both
among gods and men. It contains the bond by which
the whole and the universe are held together, and
also that by which the gods and men are connected.
Among the celestials it is called Themis, and among
the terrestrials it is called Dice; while among men

it is called the Law. These are but symbols indic-
ative that justice is the supreme virtue. Virtue,
therefore, when it consists in contemplating
and judging, is called prudence; when in sustain-
ing dreadful things, is called fortitude; when in
restraining pleasure, it is called temperance;
and when in abstaining from injuring our neighbors, justice.

Obedience to virtue according to, and transgression thereof contrary to right reason, tend
towards decorousness, and its opposite. Propriety
is that which ought to be. This requires neither
addition or detraction, being what it should be.
The improper is of two kinds: excess and defect.
The excess is over-scrupulousness, and its deficiency, laxity. Virtue however is a habit of propriety. Hence it is both a climax and a medium,
of which are proper things. They are media because they fall between excess and deficiency;
they are climaxes, because they endure neither
increase nor decrease, being just what they ought
to be.

Since however the virtue of manners consists
in dealing with the passions, over which pleasure and pain are supreme, virtue evidently does
not consist in extirpating the passions, of the
soul, pleasure and pain, but in regulating them.
Not any more does health, which is an adjustment
of the bodily powers, consist in expelling the
cold and the hot, the moist and the dry, but in
adjusting them suitably, and symmetrically. Likewise in music, concord does not consist in expelling the sharp and the flat, but in exterminating
dissonance by concord arising from their adjustment. Therefore it is the harmonious adjustment
of heat and cold, moisture and dryness which
produces health, and destroys disease. Thus by
the mutual adjustment of anger and desire, the
vices and other passions are extirpated, while
virtues and good manners are induced. Now the
greatest peculiarity of the virtue of manners
in beauty of conduct is deliberate choice. Reason and power may be used without virtue, but
deliberate choice cannot be used without it; for

deliberate choice inspires dignity of manners.

When the reasoning power by force subdues
anger and desire, it produces continence and
endurance. Again, when the reasoning force is
dethroned violently by the irrational parts,
then result incontinence and effeminacy. Such
dispositions of the soul as these are half-perf-
ect virtues and vices. For (according to its nature) the reasoning power of the soul induces
health, while the irrational induces disease.
So far as anger and desire are governed and led
by the soul's rational part, continence and
endurance become virtues; but in so far as this
is effected by violence, involuntarily, they
become vices. For virtue must carry out what is
proper not with pain but pleasure. So far as
anger and desire rule the reasoning power, there
is produced effeminacy and incontinence, which
are vices; but in so far as they gratify the
passions with pain, knowing that they are erron-
eous, in consequence of the eye of the soul being
healthy, so far as this is the case, they are
not vices. Hence it is evident that virtue must
voluntarily do what is proper, as the involuntary implies pain and fear, while the voluntary
implies pleasure and delight.

This may be corroborated by division. Knowled
and the perception of things are the province of
the rational part of the soul; while power perta
to the irrational part, whose peculiarity is ina
ility to resist pain, or to vanquish pleasure. I
both of these, the rational and the irrational
subsists deliberate choice, w hich consists of
intention and appetite, intention pertaining to
the rational part, and appetite to the irrationa
Hence every virtue consists in a mutual adaptati
of the soul's parts, while both will and deliber
ate choice subsist entirely in virtue.

In general, therefore, virtue is a mutual ad
tation of the irrational part of the soul to the
rational. Virtue, however, is produced through
pleasure and pain striking the right resultance
of propriety. But propriety is that which ought
to be, and the improper, what ought not;,,,,,,,

The fit and the unfit are to each other as the
equal and the unequal, as the ordered and the
disordered; of which the two former are finite,
and the two latter infinite (limit and infinity
are the two great principles of things, below
the universal ineffable cause). On this account
the parts of the unequal are referred to the
middle, but not to each other. An angle greater than a right angle is called obtuse; the acute
one being less than it. (In a circle) also, the
right line is greater, than the radius, drawn
from the centre. Any day beyond the equinox is
greater than it. Overheat or undercold produce
diseases. Overheatedness exceeds moderation,
which over-coldness does not reach.

This same analogy holds good in connection
with the soul. Boldness is an excess of propriety
in the endurance of things of a dreadful nature;
while timidity is a deficiency. Prodigality is
an excess of proper expenditure of money; while
illiberality is its excess. Rage is an excess of
the proper use of the soul's irascible part,
while insensibility is the corresponding deficiency. The same reasoning applies to the
opposition of the other dispositions of the
soul.

Since however virtue is a habit of propriety,
and a medium of the passions, it should be neither
wholly impassive, nor immoderately passive. Im-
passivity causes unimpelledness of the soul and
lack of enthusiasm for the beautiful in conduct,
while immoderate passivity perturbs the soul, and
makes it inconsiderate. We should then, in virtue,
see passions as shadow and outline in a picture;
which depend on animation and delicacy, imitation
of the truth and contrast of coloring. The soul's
soul's passions are animated by the natural incitation and enthusiasm of virtue, which is generated from the passions, and subsisting with
them. Similarly, harmony includes the sharp and
the falt, and mixtures consist of heat and cold,
and equilibrium results from weight and light-

ness. Therefore, neither would it be necessary
nor profitable to remove the passions of the
soul; but they must be mutually adjusted to
the rational part, under the direction of propriety and moderation.

ZALEUCUS THE LOCRIAN

PREFACE TO HIS LAWS

All inhabitants of city or country should in the
first place be firmly persuaded of the existence
of divinities, as result of their observation
of the heavens and the world, and the orderly
arrangement of their contained beings. These are
not the productions of fortune or of men. We
should reverence and honor them as causes of
every reasonable good. We should therefore prepare our souls so they may be free from vice,
For the gods are not honored by the worship of
a bad man, nor through sumptuousity of offerings, nor with the tragical expense of a depraved
man; but by virtue, and the deliberate choice of
good and beautiful deeds. All of us, therefore,
should be as good as possible, both in actions
and deliberate choice; if he wishes to be dear
to divinity; He should not fear the loss of money
more than that od renown; such a one would be considered the better citizen.

Those who do not easily feel so impelled, and
whose soul is easily excited to injustice, are
invited to consider the following. They, and
their fellow-residents of a house should remember
that there are Gods who punish the unjust, and
should remember that no one escapes the final
liberation from life. For in the supreme moment
they will repent, from remembering their unjust
deeds, and wishing that their deeds had been just.
Every one, in every action should be mindful of
this time, as if it were present; which is a powerful incentive to probity and justice.

Should any one feel (tempted by) the presence
of an evil genius, tempting him to injustice, he
should go into a temple, remain at the altar,
or in sacred groves, flying from injustice as
from an impious and harmful mistress, supplicat-
ing the divinities to cooperate with him in
turning it away from himself. He should also seek
the company of men known for their probity, in

order to hear them discourse about a blessed life
and the punishment of bad men, that he may be deterred from bad deeds, dreading none but the avenging geniuses.

Citizens should honor all the Gods according
to the particular country's legal rites, which
should be considered as the most beautiful of all
others. Citizens should, besides obeying the laws,
show their respect for the rulers by rising before
them, and obeying their instrucions. Men who are
intelligent, and wish to be saved should, after
the Gods, geniuses and heroes most honor parents,
laws, and rulers.

Let none love his city better than his country, the indignation of whose gods he would thus
be exciting; for such conduct is the beginning of
treachery. For a man to leave his country and reside in a foreign land, is something most afflict-
ing and unbearable; for nothing is more kindred to
us than our natal country. Nor let any one conside
a naturalized citizen an implacable enemy; such
a person could neither judge, nor govern properly,
for his anger predominates over his reason. Let
none speak ill either of the whole city, or of a
private citizen.

Let the guardians of the laws keep a watchful
eye over offenders, first by admonishing them, and
if that is not sufficient, by punishment. Should
any established law seem unsatisfactory, let it be
changed into a better one; but whichever remain
should be universally obeyed; for the breaking
of established laws is neither beautiful nor beneficial; though it is both beautiful and beneficial
to be restrained by a more excellent law, as if ve
vanquished thereby.

Transgressors of established laws shoul howeve
be punished, as promoting anarchy, which is the
greatest evil. The magistrates should neither be
arrogant, nor judge insultingly, nor in passing
sentence regard friendship. or hate, being partial, thus deciding more justly, and being worthy
of the magistracy. Slaves should do what is just
through fear, but free men, through shame, and fo
the sake of beauty in conduct. Governors should be
men of this kind, to arouse reverence.

Any one who wishes to change any one of the est established laws, or to introduce another law, should put a halter around his neck, and address the people. And if from the suffrages it should appear that the established law should be dissolved, or that a new law should be introduced, let him not be punished. But if it should appear that the preexisting law is better, or that the new proposition is unjust, let him who wishes to change an old, or introduce a new law, be executed by the halter.

CHARONDAS THE CATANEAN,

PREFACE TO HIS LAWS

From the Gods should begin any deliberation
or performance; for according to the old proverb
"God should be the cause of all our deliberation
and works." Further, we should abstain from base
actions especially on account of consulting with
the gods; for there is no communication between
God and the unjust.

Next, every one should help himself, inciting
himself to the undertaking and performance of su
things as are conformable to his abilities; for
it seems sordid and illiberal for a man to exten
himself similarly to small and great undertakin
You should carefully avoid rushing into things
too extensive, or of too great importance. In
every undertaking you should measure your own
desert and power, so as to succeed and gain
credit.

A man or woman condemned by the city should
not be assisted by anybody; any one who should
associate with him should be disgraced, as similar to the condemned. But it is well to love men
who have been voted approved, and to associate
with them; to imitate and acquire similar virtue
and probity, thus being initiated in the greatest and most perfect of the mysteries; for no ma
is perfect without virtue.

Assistance should be given to an injured
citizen, whether he is in his own, or in a forei
country. But let every stranger who was venerate
in his own country, and conformably to the prope
laws of that country, be received or dismissed
with auspicious cordiality, calling to mind hos-
pitable Jupiter, as a God who iis established by
all nations in common, and who is the inspective
guardian of hospitality and inhospitality.

Let the older men preside over the younger, s
that the latter may be deterred from, and ashame
of vice, through reverence and fear of the forme

For where the elders are shameless, so also are
their children and grandchildren. Shamelessness
and impudence result in insolence and injustice;
and of this the end is death.

Let none be impudent, but rather modest and
temperate; for he will thus earn the propitious-
ness of the Gods, and for himself achieve salvation; no vicious man is dear to the divinities.
Let every one honor probity and truth, hating
what is base and false. These are the indications
of virtue and vice. From their very youth children
should therefore be accustomed (to worthy manners); by punishing those who love falsehood,
and delighting those who love the truth, so as
to implant in each what is most beautiful, and
most prolific of virtue.

Each citizen should be more anxious for a
reputation for temperance than for wisdom, which
pretense often indicates ignorance of probity,
and pusillanimity. The pretense to temperance
should lead to a possession of it; for no one
should feign with his tongue that he performs
beautiful deeds, when destitute of worthiness and
good intentions.

Men should preserve kindness towards their
rulers, obeying and venerating them as if they
were parents; for whoever cannot see the propriety
of this will suffer the punishment of bad counsels from the geniuses who are the inspective
guardians of the seat of empire. Rulers are the
guardians of the city, and of the safety of the
citizens.

Governors must preside justly over their
subjects, in a manner similar to that over their
own children, in passing sentences on others,
propitiating hatred, and anger.

Praise and renown is due the rich who have assisted the indigent; they should be considered
saviors of the children and defenders of their
country. The wants of those who are poor through
bad fortune should be relieved; but not the wants
resulting from indolence or intemperance. While
fortune is common to all men, indolence and intemperance is peculiar to bad men.

Let it be considered as a worthy deed to
point out any one who has acted unjustly, in order
that the state may be saved, having many guardians of its proprieties. Let the informer be considered a pious man, though his information affect
his most familiar acquaintance; for nothing is
more intimate or kindred to a man than his country. However let not the information regard things
done through involuntary ignorance, but of such
crimes as have been committed from a previous
knowledge of their enormity. A criminal who shows
enmity to the informer should be generally hated,
that he may suffer the punishment of ingratitude,
through which he deprives himself of being cured
of the greatest of diseases, namely, injustice.

Further, let contempt of the Gods be considered as the greatest of iniquities, also voluntary
injury to parents, neglecting of rulers and laws,
and voluntary dishonoring of justice. Let him be
considered as a most just and holy citizen who
honors these things, and to the rulers indicates
the citizens that despise them.

Let it be esteemed more honorable for a man to
die for his country, than through a desire of life
to desert it, along with probity; for it is better
to die well than to live basely and disgracefully.

We should honor each of the dead not with tears
or lamentations, but with good remembrance, and
with an oblation of annual fruits. For when we
grieve immoderately for the dead, we are ungrate-
ful to the terrestrial geniuses.

Let no one curse him by whom he has been injured; praise is more divine than defamation.

He who is superior to anger should be considered a better citizen than he who therethrough
offends.

Not praiseworthy, but shameful is it to
surpass temples and palaces in the sumptuousness
of his expense. Nothing private should be more
magnificent and venerable than things of a public
nature.

Let him who is a slave to wealth and money
be despised, as pusillanimous and illiberal, being

impres sed by sumptuous possessions, yet leading
a tragical and vile life. The magnanimous man
foresees all human concerns, and is not disturbed
by any accident of fortune.

Let no one speak obscenely; lest his thoughts
lead him to base deeds, and defile his soul with
impudence. Proper and lovely things it is well
and legal to advertise; but such things are honored
by being kept silent. It is base even to mention
something disgraceful.

Let every one dearly love his lawful wife,
and beget children by her. But let none shed
the seed due his children into any other person,
and let him not disgrace that which is honorable
by both nature and law. For nature produced the
seed for the sake of producing the children, and
not for the sake of lust.

A wife should be chaste, and refuse impious
connection with other men, as by so doing she will
will subject herself to the vengeance of the
geniuses, whose office it is to expel those to
whom they are hostile from their houses, and to
produce hatred.

He who gives a step-mother to his children
should not be praised, but disgraced, as the
cause of domestic dissension.

As it is proper to observe these mandates, let
him who transgresses them be subjected to political execration.

The law also orders that these introductory
suggestions be known by all the citizens, and
should be read in the festivals after the hymns
to Apollo called paeans, by him who is appointed
for this purpose by the master of the feast, so
that these precepts may germinate in the minds
of all who hear them.

CALLICRATIDAS

ON THE FELICITY OF FAMILIES

The universe must be considered as a system of kindred communion or association. But every system consists of certain dissimilar contraries, and is organized with reference to one particular thing, which is the most excellent, and also with a view to benefit the majority. What we call a choir is a system of musical communion in view of one common thing, a concert of voices. Further a ship's construction-plan contains many dissimilar contrary things, which are arranged with reference to one thing which is best, the pilot; and the common advantage of a prosperous voyage.

Now a family is also a system of kindred communion, consisting of dissimilar proper parts; organized in view of the best thing, the father of the family, the common advantage being unanimity. In the same manner as a zither, every family requires three things, apparatus, organization, and a certain manner of practise, or musical use. An apparatus being the composition of all its parts, is that from which the whole, and the whole system of kindred communion derives its consummation. A family is divided into two divisions; man and the possessions; which latter is the thing governed, that affords utility. Thus also, an animal's first and greatest parts are soul and body; soul being that which governs and uses, the body being that which is governed, and affords utility. Possessions indeed are the adventitous instruments of human life, while the body is a tool born along with the soul, and kindred to it. Of the persons that complete a family, some are relatives, and others only attracted acquaintances. The kindred are born from the same blood, or race. The affinities are an accidental alliance, commencing with the communion of wedlock. These are either fathers or brothers, or maternal and paternal grandfathers, or other relatives by marriage.

But if the good arising from friendship is also
to be referred to a family, — for thus it will
become greater and more magnificent, not only
through an abundance of wealth and many relations,
but also through numerous friends, — in this case
it is evident that the family will thus become more
more ample, and that friendship is a social rela-tion
tion essential to a family. Possessions are either necessary or desirable. The necessary subserve
the wants of life; the desirable produce an elegant and well-ordered life, replacing many other
znecessaries?). However, whatever exceeds what is n
is needed for an elegant and well-ordered life
are the roots of wantonness, insolence and destruction. Great possessions swell out with pride,
and this leads to arrogance, and fastidiousness, cc
conceiving that their kindred, nation and tribe
do not equal them. Fastidiousness leads to insolence, whose end is destruction. Wherever then, in f
in family or city there is a superfluity of possessions, the legislator must cut off and amputate
ate the superfluities, as a good husbandman prunes
luxurious leafage.

In the family's domestic part there are three
divisions: the governor, the husband; the governed,
the wife; and the auxiliary, the offspring.

- - - - - - - - -

With respect to practical and rational domination, one kind is despotic, another protective,
and another political. The despotic is that which g
governs with a view to the advantage of the governor, and not of the governed, as a master rules
his slaves, or a tyrant his subjects. But the
guardian domination subsists for the sake of the
governed, and not the governor; as the masseurs
rule the athletes, physicians over the sick, and
preceptors over their pupils. Their labors are
not directed to their own advantage, but to the
benefit of those they govern; those of the phys-
ician being undertaken for the sake of the sick,
that of the masseurs for the sake of exercising
somebody else's body, and those of the erudite

for the ignorant. Political domination, however,
aims at the common benefit of both governors and
governed. For in human affairs, according to this
domination, are organized both a family and a
city: just as the world and divine affairs are
in correspondence. A family and a city stand in
a relation analogous to the government of the
world. Divinity indeed is the principle of nature, and his attention is directed neither to
his own advantage, nor to private good, but to
that of the public. That is why the world is called cosmos, from the orderly disposition of all
things, which are mutually organized of the most
excellent thing, which is God, who, according
to our notions of him, is a celestial living
being, incorruptible, and the principle and cause
of the orderly disposition of the wholes.

Since therefore the husband rules over the
wife, he rules with a power either despotic, protective, or political. Despotic power is out of
the question, as he diligently attends to her
welfare; nor is it protective entirely, for he has
to consider himself also. It remains therefore
that he rules over her with a political power,
according to which both the governor and governed
seek the common advantage. Hence wedlock is established with a view to the communion of life.
Those husbands that govern their wives despotic-
ally are by them hated; those that govern them
protectively are despised; being as it were mere
appendages and flatterers of their wives. But
those that govern them politically are both admired and beloved. Both these will be effected
if he who governs exercises his power so that
it may be mingled with pleasure and veneration;
pleasure at his fondness, but veneration at his
doing nothing vile or abject.

- - - - - - - - - - - - - -

He who wishes to marry ought to take for a
wife one whose fortune is conformable to his own,
neither above nor beneath, but of equal property.

Those who marry a woman above their condition have
to contend for the mastership; for the wife, surpassing her husband in wealth and lineage, wishes
to rule over him; but he considers it to be worthy
of him, and unnatural to submit to his wife. But
those who marry a wife beneath their condition
subvert the dignity and reputation of their family. One should imitate the musician, who having
learned the proper tone of his voice, moderates
it so as to be neither sharp nor flat, nor broken,
nor strident. So wedlock should be adjusted to
the tone of the soul, so that the husband and wife
may accord, not only in prosperity, but also in
adversity. The husband should be his wife's regulator, master and preceptor. Regulator, in paying diligent attention to his wife's affairs;
master, in governing and exercising authority
over her, and preceptor in teaching her such
things as are fitting for her to know. This will
be specially effected by him who, directing his
attention to worthy parents, from their family
marries a virgin in the flower of her youth. Such
virgins are easily fashioned, and docile; and are
naturally well disposed to be instructed by, and
to fear and love their husbands.

PERICTYONE

ON THE DUTIES OF A WOMAN

A woman should be a harmony of prudence and temperance. Her soul should be zealous to acquire virtue; so that she may be just, brave, prudent, frugal, and hating vain-glory. Furnished with these virtues, she will, when she becomes a wife, act worthily towards herself, her husband, her children and her family. Frequently also such a woman will act beautifully towards cities, if she happens to rule over cities and nations, as we see is sometimes the case in a kingdom. If she subdues desire and anger, there will be produced a divine symphony. She will not be pursued by illegal loves, being devoted to her husband, children and family. Women fond of connections with outside man come to hate their families, both the free members, and the slaves. They also plot against their husbands, falsely representing them as the calumniators of all their acquaintance, so that they alone may appear benevolent; and they govern their families in a way such as may be expected from lovers of indolence. Such conduct leads to the destruction of everything common to husband and wife.

The body should also be trained to moderatio in food, clothes, baths, massage, hair-dressing, and jewelry adornment. Sumptuous eating, drinking, garments and keepsakes involve them in every crime, and faithlessness to their husband and everybody else. It is sufficient to satisfy hunger and thirst, and this from easily accessible things; and protect themselves from the cold by garments of the simplest description. It is quite a vice to feed on things brought from distant countries, and bought at a great price. It is also great folly to search after excessively elegant garments, made brilliant wi with purple or other precious colors.

The body itself demands no more than to be saved from cold and nakedness, for the sake

propriety, and that is all it needs. Men's opinions, combined with ignorance, demands inanities
and superfluities. No woman should be decorated
with gold, nor gems from India nor any other
country, nor plait her hair artistically, nor be
perfumed with Arabian perfumes, nor paint her
face so that it may be more white or more red,
nor give a dark tinge to her eyebrows and her eyes,
nor artificially dye her gray hair, nor bathe
vontinually. A woman of this sort is hunting a
spectator of female intemperance. The beauty produced by prudence, and not by these particulars,
pleases women that are well born. Neither should
she consider it necessary to be noble, rich, birth-
in a great city, glory, have glory, and the friende
ship of renowned or royal men. The presence of
such should not cause her any annoyance, but should
they be absent, she should not regret them; their
absence will not hinder the prudent woman from
living properly. Her soul should not anxiously
dream about them, but ignore them. They are really more harmful than beneficial, as they mislead
to misfortune; inevitable are treachery, envy and
calumny, so that their possessor cannot be free
from perturbation.

She should venerate the Gods, thereby hoping
to achieve felicity, also by obeying the laws and
sacred institutions of her country. After the gods,
she should honor and venerate her parents, who
cooperate with the gods in benefiting their chip-
dren.

Moreover she ought to live with her husband
legally and kindly, claiming nothing as her own
property, but preserving and protecting his bed;
for this protection contains all things. In a becoming manner she should bear any stroke of fortune that may strike her husband; whether he is unfortunate in business, or makes ignorant mistakes,
is sick, intoxicated, or has connection with other
women. This last is a privilege granted to men, but
but not to women, since they are punished for this
offence. She must submit to the law with equanimity, without jealousy. She should likewise patiently

bear his anger, his parsimony, complaints he may
make of his destiny, his jealousy, his accusations
of her, and whatever other faults he may inherit
from his nature. All these she should cheerfully
endure, conducting herself towards him with prudence and modesty. A wife who is dear to her husband, and who truly performs her duty towards him,
is a domestic harmony, and loves the whole of her
family, to which also she conciliates the benevolence of strangers.

If however she loves neither her husband nor
her children, nor her servants, nor wishes to see
any sacrifice preserved, then she becomes the herald of every kind of destruction, which she likewise prays for, as being an enemy, and also prays
for the death of her husband, as being hostile to
him, in order that she may be connected with other
men; and in the last place she hates whatever
her husband loves.

But a wife will be a domestic harmony if she
is full of prudence and modesty. F or then she
will love not only her husband, but also her children, her kindred, her servants, and the whole of
her family, among which she numbers her possessions, friends, fellow-citizens, and strangers.
Their bodies she will adorn without any superfluous ornaments, and will both speak and hear such
things only as are beautiful and good. She should
conform to her husband's opinion in respect to
their common life, and be satisfied with those relatives and friends as meet his approbation. Unless
she is entirely devoid of harmony she will consider pleasant or disagreeable such things which are
thought so by her husband.

ON THE HARMONY OF A WOMAN

Parents ought not to be injured either in
word or deed; and whatever their rank in life,
small or great, they should be obeyed. Children
should remain with them, and never forsake them,
and almost to submit to them, even when they are
insane, in every allotted condition of soul or bod

body, or external circumstances, in peace, war,
health, sickness, riches, poverty, renown, ignominy, class, or magistrate's rank. Such conduct
will be wisely and cheerfully adopted by the pious
pious. He who despises his parents will both among
the living and the dead be condemned for this crime
by the Gods, will be hated by men, and under earth
will, together with the impious, be eternally punished in the same place by Justice, and the subterranean Gods, whose province it is to inspect things
of this kind.

The aspect of pa rents is a thing divine and
beautiful, and a diligent observance of them is
attended by a delight such that neither a view of
the sun, nor of all the stars, which swing around
the illuminated heavens, is capable of producing
any spectacle greater than this. The Gods are not
envious in a case like this.

We shoulld reverence parents both while living
and dead, and never oppose them in any thing they
say or do. If ignorant of anything through de-
ception or disease, their children should console
and instruct, but by no means hate them on this
account. For no greater error or injustice can be
committed by men than to act impiously towards
their parents.

ARISTOXENUS OF TARENTUM

APOTHEGMS.

After divinity and geniuses, the greatest respect should be paid to parents and the laws; not fictitiously, but in reality preparing ourselves to an observance of, and perseverance in, the manners and laws of our country, though they should be in a small degree worse than those of other countries.

(FROM THE FOURTH BOOK).

But after these things follow the honors which should be paid to living parents, it being right to discharge the first, the greatest, and the most ancient of all debts. Every one, likewise, should think that all which he possesses belongs to those who begot and nurtured him, in order that he may be ministrant to their want to the utmost of his ability, beginning from his property; in the second place, discharging his debt to them from things pertaining to his body; and in the third place, from things pertaining to his soul; thus with usury repaing the cares and pains which his now very aged parents bestowed on him when he was young. Through the whole of life, likewise, he should particularly employ the most respectful language in speaking to his parents; because there is a most severe punishment for light and winged words; and Nemesis, the messenger of Justice, is appointed to be the inspector of everything of this kind.

When parents a re angry, therefore, we should yield to them, and appease their anger, whether it is seen inwords or deeds; acknowledging that a father may reasonably be very much enraged with his son, when he thinks that he has been injured by him.

On the parents' death, the most appropriate and beautiful monuments should be raised to them; not exceeding the usual magnitude, nor yet less

than those which our ancestors erected for their
parents. Every year, also, attention ought to be
paid to the decoration of their tombs. They should
likewise be continually remembered and reverenced,
and this with a moderate but appropriate expense.

  By always acting and living in this manner
we shall each of us be rewarded according to our
deserts, both by those Gods and those natures that
are superior to us, and shall pass the greatest par-
part of our life in good hope.

EURYPHAMUS

CONCERNING HUMAN LIFE.

The perfect life of man falls short indeed of the life of God, because it is not self-perfect, but surpasses that of irrational animals, participating as it does of virtue and felicity. For neither is God in want of external causes, — as he is naturally good and happy, and is perfect from himself; — nor any irrational animal. For brutes being destitute of reason, they are also destitute of the sciences pertaining to actions. But the nature of man partly consists of his own proper deliberate choice, and partly is in want of the assistance derived from divinity. For that which is capable of being fashioned by reason, which has an intellectual perception of things beautiful and base, can from earth erect itself and look to heaven, and with the eye of intellect can perceive the highest Gods, — that which is capable of all this likewise receives assistance from the Gods.

But in consequence of possessing will, deliberate choice, and a principle of such a kind as enables it to study virtue, and to be agitated by the storms of vice, to follow, and also to apostacize from the Gods, — it is likewise able to be moved by itself. Hence it may be praised or blamed, partly by the Gods, and partly by men, according as it applies itself zealously either to virtue or vice.

For the whole reason of the thing is as follows: Divinity introduced man into the world as a most exquisite being, to be honored reciprocally with Divinity, and as the eye of the orderly systematization of everything. Hence also man gave things names, himself becoming the character of them. He also invented letters, through these procuring a treasury of memory. He imitated the established order of the universe, by laws and judicial proceedings organizing the communion of cities. For no human work is more honorable in the eyes of the world, nor more worthy of notice by the Gods, than

proper constitution of a city governed by good laws,
distributed in an orderly fashion throughout the
state. For though by himself no man amounts to
anything, and by himself is not able to lead a
life conforming to the common concord, and to
the proper organization of a state, yet he is well
adapted to the perfect system of society.

Human life resembles a properly tuned and
cared for lyre. Every lyre requires three things:
apparatus, tuning, and musical skill of the player.
By apparatus we mean preparation of all the appropriate parts; the strings, (the plectrum) and other
instruments cooperating in the tuning of the
instrument. By tuning we mean the adaptation
of the sounds to each other. The musical skill is
the motion of the player in consideration of the
tuning. Human life requires the same three things.
Apparatus is the preparation of the physical basis
of life, riches, renown, and friends. Tuning is the
organizing of these according to virtue and the
laws. Musical skill is the mingling of these according to virtue and the laws, virtue sailing with a
prosperous wind, with no external resistance. For
felicity does not consist in being driven from the
purpose of voluntary intentions, but in obtaining
the; nor in virtue lacking attendants and servers;
but in completely possessing its own proper powers
which are adapted to actions.

For man is not self-perfect; he is imperfect.
He may become perfect partly from himself, and
partly from some external cause. Likewise, he may
be perfect either according to nature or to life.
According to nature he is perfect, if he becomes
a good man; as the virtue of everything is the
climax and perfection of the nature of that thing.
Thus the virtue of the eyes is the climax and
perfection of their nature; and this is also true
of the virtue of the ears. Thus too the virtue of
man is the climax and perfection of the nature of
man. But man is perfect according to life when he
becomes happy. For felicity is the perfection and
completion of human goods. Hence, again, virtue
and prosperity become parts of the life of man.

Virtue, indeed, is a part of him so far as he is soul; but prosperity, so far as he is connected with body; but both parts of him, so far as he is an animal. For it is the province of virtue to use in a becoming manner the goods which are conformable to nature; but of prosperity to impart the use of them. The former, indeed imparts deliberate choice and right reason; but the latter, energies and actions. For to wish what is beautiful in conduct, and to endure things of a dreadful nature, is the proper business of virtue. But it is the work of prosperity to render deliberate choice successful, and to cause actions to arrive at the desired end. For a general conquers in conjunction with virtue and good fortune. The pilot sails well in conjunction with art and prosperous winds; the eye sees well in conjunction with acuteness of vision, and light. So the life of man reaches its perfection through virtue itself, and prosperity.

HIPPARCHUS

ON TRANQUILLITY

Since men live but for a very short period, if their life is compared to the whole of time, they will, as it were, make a most beautiful journey, if they pass through life with tranquility. This they will best possess if they will accurately and scientifically know themselves, namely that they are mortal and of a fleshly nature, and that they have a body which is corruptible, and can be easily injured, and which is exposed to everything most grievous and severe, even to their latest breath.

In the first place, let us observe those things which happen to the body; such as pleurisy, pneumonia, phrensy, gout, strangury, dysentery lethargy, epilepsy, ulcers, and a thousand other diseases. But the diseases that can happen to the soul are much greater and direr. For all the iniquitous, evil, lawless and impious conduct in the life of man, originates from the passions of the soul. For through unnatural immoderate desires many have become subject to unrestrained impulses; and have not refrained from the most unholy pleasures, arising from connections with daughters and even mothers. Many have even destroyed their fathers and offspring. But what is the use to continue detailing externally impending evils, such as excessive rain, draught, violent heat, and cold; so that frequently from the anomalous state of the air, pestilence and famine arise, followed by manifold calamities making whole cities desolate. Since therefore many such calamities impend, we should neither be elated by the possession of worldly goods, which might rapidly be consumed by the irruption of some small fever, nor with what are conceived to be prosperous external circumstances, which from their own nature frequently decay quicker than they arose. For all these are uncertain and un-

stable, and are found to have their existence in
many and various mutations; and no one of them is
permanent, or immutable, or stable, or indivisible.
Consideringn these things well, and also being
persuaded that if what is present and is imparted
to us, is able to remain for the smallest portion
of time, it is as much as we ought to expect; we
shall then live in tranquility, and with hilarity,
generously bearing whatever may befall us.

How many people imagine that all they have
and what they receive from fortune and nature
is better than it is, not realizing what it is in
reality; but such as it is able to become when
it has arrived at its highest excellence, they then
burden the soul with many and great, and nefarious
stupid evils, when they are sudrenly deprived of
these transitory goods. That is how they lead a
most bitter and miserable life. But this takes
place in the loss of riches, os the death of frion
friends and children, or in the privation of certain other things, which by them are conceived to
be possessiond most honorable. Afterwards, weeping and lamenting, they assert of themselves, that
they alone are most unfortunate and miserable,
not remembering that these things have happened,
and even now happen to many others; nor are they
able to understand the life of those that are now
in existence, and of those that have lived in
former times, nor to see in what great calamities
and waves of evils many of the present times are,
and of the past have been involved. Therefore cons
sidering with ourselves that many who have lost
their property have afterwards on account of this
very loss been saved, since thereafter they might
either have fallen into the hands of robbers, or
into the power of a tyrant; that many also who ha
loved certain persons, and have been extremely be
evolently disposed towards them, but have afterwa
hated them extremely, — considering all these
things, of which history informs us; and learning
likewise that many have been destroyed by their
own children, and by those they have most dearly
loved, and comparing our own life with that of

THOSE WHO HAVE BEEN MORE UNHAPPY THAN WE have
been, and taking into account general human vics
issitudes, that happen to others beside ourselves, we shall pass through life with greater
tranquility.

A reasonable man wil l not think the calam -
ities of others easy to be born, but not his own;
since he sees that the whole of life is naturally
exposed to many calamities. Those however who
weep and lament, besides not being able to re-
cover what they have lost, or recall to life
those that are dead, impel the soul to still
greater perturbations; in consequence of its being
filled with much depravity. Being washed and purified, we should do our best to wipe away our
inveterate stains, by the reasonongs of philosophy. This we shall accomplish by adhering to
prudence and temperance, being staisfied with
our present circumstances, and not aspiring after too many things. Men who gather a great
abundance of external things do not consider
that enjoyment of them terminates with this present life. We should therefore use the present
goods; and by the assistance of the beautiful
and venerable results of philosophy we shall be
liberated from the insatiable desire of depraved
possessions.

METOPUS

CONCERNING VIRTUE

Man's virtue is thw perfection of his nature.
By the proper nature of his virtue, every being
becomes parfect, and arrives at the cimax of its
excellence. Thus the virtue of the horse is that
which makes the best of the horse's nature. The
same reasoning applies to details. Thus the virtue
of the eyes is acuteness of vision; and this is
the climax of the eyes' nature. The virtue of the
ears is acuteness of hearing; and this is the aure
nature's climax. The virtue of the feet is swiftness; and this is the pedal nature's climax.

Every virtue, however, should include these
three things: resaon, power, and deliberate choic.
reason indeed judges and contemplates; power pro-
hibits and vanquishes; and deliberate choice love,
and enjoys propriety. Therefore to judge and contemplate pertain to the intellectual part of the
soul; to prohibit and vanquish are the peculiarity
of the irrational part of the soul; and to love
and enjoy propriety includes both rational and irrational parts of the soul: for deliberate choice
consists of the discursive energy of reason, and
appetite. Intentiona therefore,—pertains to the
rational, but appetite to the irrational parts of
thesoul.

We may discrn the multitude of the virtues
by observing theparts of thesoul; also the growth
and nature of virtue. Of the soul's parts, two
rank first: the rational and the irrational. It is
by ther rational that we judge and contemplate;
by the irrational we   . are
impelled and desire. inese are either congordant
or discordant, their strife and dissonance being
produced be excess or defect. The rational part's
victory over the irrational produces endurance
and continence; When the rational leads, the irra
rational follows, both accord, and produce virtue
That is why endurance and continence are generall
accompanied by pain; for endurance resists painl
and continence pleasure. However, incontinence

and effeminacy neither resist nor vanquish pleasure. That is why men fly form good through pain, but reject it thorough pleasure. Likewise praise and blame, and everything beautiful in human conduct, are produced in these parts of the soul. This explains the nature of virtue.

Let us study virtue's kinds and parts. Since the soul is divided into two parts, the rational and the irrational, the latter is also divided into two, the irascible and appetitive part. By the rational we judge and contemplate; by the irrational we are impelled and desire. The irascible part defends us, and revenges incidental molestations; the appetitive directs and preserves the body's proper constitution. So we see that the numerous virtues with all their differences and peculiarities do little more than conform to the distinctive parts of the soul.

CRITO

ON PRUDENCE AND PROSPERITY

Such is the mutual relation of prudence and
prosperity. Prudence is explainable and reasonable,
orderly and definite. Prosperity is inexplainable,
and irrational, disorderly, and indefinite. In
origination and power prudence is prior to prosperity; the former governing and defining, the
latter being governed and defined; but they are
mutually adjusting, concurring in the same thing.
For that which limits and adjusts must be explainable and reasonable; while that which is limited
and adjusted is naturally unexplainable and irrational. That is how the reason of the infinite's
nature, and of the limiter subsists in all things.
Infinites are always naturally disposed to be limited and adjusted by things possessing reason and
prudence; for in relation to the latter the former
stand as matter and essence. But finites are self-
adjusted and self-limited, being causal and energe-
etic.

The mutual adjustment of these natures in
different things produces a variety of adjusted
substances. For in the comprehension of the whole
of things, the mutual adjustment of both the moving and the passive, is the world. There is no
other possible way of salvation for the whole and
the universe, than by the adjustment of the things
generated to the divine, and of the ever-passive
to the ever moved. The similar adjustment, in man,
of the irrational to the rational part of the soul
is virtue. for this cannot exist in case of mutual
strife between the two. So also in a city, the
mutual adjustment of the governors to the governed
produces strength and concord. Governing is the
specialty of the better nature; while being governed is more suited to the subordinate part.
To both are common strength and concord. A simila:
mutual adjustment exists in the universe and in
the family; the latter being a resultance of
allurements and erudition with reason, the father
of pains and pleasures, prosperity and adversity.

Man's constitution is such that he needs changes. work and rest, sorrow and gladness, prosperity and adversity. Somethings draw the intellect towards wisdom, and industry, and keep it there; others relax and delight, rendering the intellect vigorous and prompt. Should one of these elements prevail, then man's life becomes one-sided, exaggerating sorrow. and difficulty, or levity and smoothness. Now all these should be mutually adjusted by prudence, which discerns and distinguishes in actions the elements of limitation and infinity. That is why prudence is the mother and leader of the other virtues. For it is prudence's reason and law which organize and harmonize all other virtues.

Summarizing: The irrational and explainable are to be found in all things; the latter defines and limits, the former is defined and bounded. The resultance of both is the proper organization of the whole and the universe.

— — — — — — —

God fashioned man in a way such as to declare that not through the want of power or deliberate choice, that man is incapable of impulsion to beauty of conduct. In man was implanted a principle such as to combine the possible with the desirable; so that while man is the cause of power and of the possession of good, God is that of reasonable impulse and incitation. So God made man tend to heaven, gave him an intellective power, implanted in him a sight called intellect, which is capable of beholdingg God. For without God, it is impossible to discover what is best and most beautiful; and without intellect we cannot see God, since every mortal nature's establishment implied a progressive loss of intellect. It is not God, however, who effected this, but generation, and that impulse of the soul which lacks deliberate choice.

POLUS

ON JUSTICE

I think that the justice which subsists among men may be called the mother and nurse of the other virtues. Without it no man can be temperate, brave, nor prudent. In conjunction with elegance it is the harmony and peace of the whole soul. This virtue's strength will become more manifest if we compare it to the other habits. They have a partial utility, and refer to one thing only; while this refers to a multitude, nay, to whole systems. It conducts the whole world-government, and is called providence, harmony, and Vengeance (Dike), by the decrees of a certain kind of geniuses. In a city it is justly called peace, and equitable legislation. In a house, it is the concord between husband and wife; the kindliness of the servant towards his master, and the anxious care of the master for his servant. In the body, likewise, which to all animals is the first and dearest thing, it is the health and wholeness of each part. In the soul it is the wisdom that depends from science and justice. As therefore this virtue disciplines and saves both the whole and parts of everything, mutually tuning and familiarizing all things, it surely deserves, by universal suffrages, to be called the mother and nurse of all things.

STHENIDAS THE LOCRIAN

ON A KINGDOM

A king should be a wise man; thus will he be honored in the same manner as the supreme divinity, whose imitator he will be. As the Supreme is by nature the first King and potentate, so will a king be, by birth and imitation. As the former rules in the universe, and in the whole of things, so does the latter in the earth. While the former governs all things eternally, and has a never-failing life, possessing all wisdom in himself, so the latter acquires science through time. But a king will imitate the First God in the most excellent manner, if he acquires magnanimity, gravity, and the restriction of his wants to but few things, to his subject exhibiting a paternal disposition.

For it is because of this especially that the First God is called the father of both Gods and men, because he is mild to everything that is subject to him, and never ceases to govern with providential regard. Nor is he satisfied with being the Maker of all things, but he is the nourisher and preceptor of everything beautiful, and the legislator to all things equally. Such also ought to be a king who on earth rules over men.

Nothing is beautiful, that lacks a director, or ruler. Again, no king or ruler can exist without wisdom and science. He therefore who is both a sage and a king will be an imitator and legitimate minister of God.

ECPHANTUS THE CROTONIAN

ON KINGS

Many arguments apparently prove that every
being's nature is adapted to the world and the
things it contains. Every animal thus conspiring
(into union and consent) and having such an organization of its parts, it follows, through the
attractive progress of the universe around it, an
excellent and necessary evolution which produces
the general ornamentation of the world, and the
peculiar permanence of everything it contains.
Hence it is called the (ornamental) kosmos,
and is the most perfect being.

When we study its parts, we find them many,
and naturally different. First, ######### a being
who is the best, b oth from its native alliance
to the world, and in its particular divinity
(containing the stars called planets, forming the
first and greatest series). Secind is the nature
of the geniuses, in the sublunary region, where
bodies move in a right line. Third, in the earth,
and with us, the best being is man, Of whom the
divinest is a king, surpassing other men in his
general being. While his body resembles that of
other men, being made of the same physical matter,
he was molded by the best sculptors, who used him
as the archetype. Hence, in a certain respect, a
king is one and alone; being the production of the
supernal king, with whom he is always familiar;
being beheld by his subjects in his kingdom as
in a splendid light.

A kingdom has been said to resemble an eagle,
the most excellent of winged animals, who undazzle
stares at the sun. A kingdom is also similar to
the sun, because it is divine; and because of its
exceeding splendor cannot be seen without difficulty, except by piercing eyes, that are genuine.
For the numerous splendors that surround it, and
the black eye-clouds it produces in those that gaz
at it, as if they had ascended into some foreign
altitude, demonstrates that their eyes are spuriou

Those however who can safely arrive thither, either
because of their familiarity therewith, or their
alliance with it, can use it properly.
    A kingdom, therefore, is something pure, genuine,
uncorrupted, and because of its preeminence, divine, and difficult of access. He who is established therein should naturally be most pure and
(think) clearly, that by his personal stains he may
may not obscure so splendid an institution; as
some persons defile the most sacdred places, and
the impure pollute those they meet. But a king,
who associates with the (best), should be undefiled,
realizing how much diviner than other things are
both himself and his prerogatives; and from the
divine exemplar of which he is an image, he should
treat both himself and his subjects worthily.
    When other men are delinquents, their most
holy purification causes them to imitate their
rulers, whether laws or king. But kings who cannot on earth find anything better than their own
nature to imitate, should not waste time in seeking
any model other or lower than God himself. No one
would long search for the world, seeing that he
exists in it, as a part of it; so the govemor
of others should not ignore him by whom he also
is governed. Being ruled is the supreme ornament,
inasmuch as there is nothing rulerless in the
universe.
    A king's manners should also be the inspiration
of his government. Thus its beauty will immediately
shine forth, since he who imitates God through
virtue will surely be dear to him whom he imitates;
and much more dear will he be to his subjects.
No one who is beloved by the divinity will be hated
by men; since neither do the stars, nor the whole
world hate God. For if they hated their ruler and
leader, they would never obey him. But it is because he governs properly that human affairs are
properly governed. The earthly king, therefore,
should not be deficient in any of the virtues
distinctive of the heavenly ruler.
    Now as an earthly king is something foreign
and external, inasmuch as he descends to men from
the heavens, so likewise his virtues may be con-

sidered as work of God, and to descend upon h
him from divinity. You will find this true, if
you study out the whole thing from the beginning.

An earthly king obtains possession of his subje
jects by an agreement, which is the first essential.
The truth of this may be gathered from the state
of affairs produced by the destruction of the usua
unanimity among citizens, which indeed is much
inferior to a divine and royal nature. Such
natures are not oppressed by any such poverty;
but, conforming to intellect, they supply the wants
wants of others, assisting them in common, being
perfect in virtue. But the friendship obtaining
in a city, and which possesses a certain common
end, imitates the concord of the universe. No city
could be inhabited without an institution of magistrates. To effect this, however, and to preserve the city, there is a necessity of laws, a polit
ical domination, and a governor and the governed.
All this happens for the general good, for unanimity, and the condent of the people in harmony
with organic efficiency. Likewise, he who governs
according to virtue, is called a king, and is so
in reality; since he possesses the same friendshi
and communion with his subjects, as divinity pos
sesses with the world, and its contained natures.
All benevolence, however, ought to be exerted, in
the first place, indeed, bybthe king towards his
subjects; second, by the subjects towards the
king; and this benevolence should be similar to
that of a parent towards his child, of a shepherd
towards his flock, and of the law towards the
law-abiding.

For there is one virtue pertaining to the
government, and to the life of men. No one should
through indigence, solicit the assistance of others, when he is able to supply himself with what
nature   requires. Though (in the city)
there is a certain community of goods, yet every
one should live so as to be self-sufficient; as
the latter requires the aid of none others in
his passgae through life. If therefore it is necessary to lead an active life, it is evident

that a king, though he should also consume other
things, will nevertheless be self-sufficient. For
have friends through his own virtue; and in using
these, he will not use them by any virtue other
than that by which he regulates his own life. For
he must follow a virtue of this kind, since he cannot procure anything more excellent. God, indeed,
needing neither ministers nor servants, nor employing any mandate, and neither crowning nor proclaiming those that are obedient to him, or disgracing
those that are disobedient, thus administers so
great an empire. In a manner to me appearing most
worthy of imitation, into all things he instills a
most zealous desire to participate in his nature.
As he is good, the most easy possible communica-
btion thereof is his only work. Those who imitate
him, find that this imitation enables them to accomplish everything else better. Indeed this imitation of God is the self-sufficiency of everything
else; for there is an identity, and no difference
between the virtues that make things acceptable
to God, and those that imitate him; and is not our
earthly king, in a similar manner self-sufficient?
By assimilating himself to one, and that the most
excellent nature, he will beneficently endeavor to
assimilate all his subjects to himself.

Such kings, however, as towards their subjects
use violence and compulsion entirely destroy in
every individual of the community a readiness to
imitate him. Without benevolence, no assimilation
is possible; since benevolence particularly effaces
fear. It is indeed much to be desired that human
nature should not be in want of persuasion; which
is the relic of human depravity, of which the tempe-
oral being called man is not destitute. Persuasion,
indeed, is akin to necessity; inasmuch as it is
chiefly used on persons flying from necessity. But
persuasion is needless with beings such as sponta-
aneously seek the beautiful and good;

Again, a king alone is capable of effecting
this human perfection, that through imitation of
the good man may pursue propriety and loveliness;
and that those who are corrupted as if by intoxication, and who have fallen into an ignorance of

the good by abad education, may be strengthened
by the king's eloquence, may have their diseased
minds healed, and their depravity's dazedness
expelled, may become mindful of an intimate associate, whose influence may persuade them.
Though originating fromu undesirable seeds, yet
(this royal influence) is the source of a certain
good to humans, in which language supplies our
deficiencies, in our mutual converse.

-- -- -- -- -- -- -- --

Hr who has a sacred and divine conception of thing
will in reality be a king. Persuaded by this, he
will be the cause of all good, but of no evil.
Evidently, as he is fitted for society, he will
become just. For communion or association consists in equality, and in its distribution. Justice
indeed precedes, but communion participates. For
it is impossible for a man to be unjust, and yet
distribute equality; or that ue should distribute
equality, and yet not be adapted to association.

How is it possible that he who is self-sufficient should not be continent? For sumptuousness is the mother of incontinence, and this of
wnton insolence, anf from this an innumerable
host of ills. But self-sufficiency is not mastered by sumptuousness, nor by any of its derivat-
ive evils, but itself being a principle, it leads
all things, and is not led by any. To govern is
the province of God, and also of a king, (on which
account indeed, he is called self-sufficient); so
to both it pertains not to be governed by any one.

Evidently, these things cannot be effected
without prudence, and it is manifest that the
world's intellectual prudence is God. For the wor
reveals graceful design, which would be impossibl
without prudence. Nor is it possible for a king
without prudence to possess these virtues; I mean
justice, continence, sociability and kindred
virtues.

PERICLUS

ON PARENTS

Neither divinity, nor anyone possessing the least
wisdom will ever advise any one to neglect his parents. Hence we cannot have any statue or temple
which will be considered by divinity as more precious than our fathers and grandfathers when grown
feeble with age. For he who honors his parents by
gifts will be recompensed by God; for without this,
divinity will not pay any attention to the prayers
of such parents for their children. Our parents' and
progenitors' images should by us be considered much
more venerable and divine than any inanimate images.
For our parents, who are divine images that are animated, when they are continually adorned and
worthily honored by us, pray for us, and implore the
Gods to bestow on us the most excellent gifts; and
do the contrary when we despise them; neither of which
which occurs with inanimate images. Hence
he who behaves worthily towards his parents and progenitors, and other kindred, will possess the
most worthy of all statues, and the best calculated
to endear him to divinity. Every intelligent person,
therefore, should honor and venerate his parents,
and should dread their execrations and unfavorable
prayers, knowing that many of them take effect.

Nature having disposed the matter thus, prudent
and modest men will consider their living aged progenitors a treasure, to the extremity of life; and
if they die before the children have arrived there,
the latter will be longing for them. Moreover, pro-

genitors will be terrible in the extreme to their

depraved or stupid offspring. The profane person who
is deaf to these considerations will by all intelligent persons be considered as odious to both
Gods and men.

PHYNTIS, DAUGHTER OF CALLICRATES

ON WOMAN'S TEMPERANCE

A woman ought to be wholly good and modest;
but she will never be a character of this kind
without virtue, which renders precious whatever
contains it. The eyes's virtue is sight; the
ears,' hearing. A horse's virtue makes it good;
while the virtue if man or woman makes them worthy. A woman's principal virtue is temperance,
wherethrough she will be able to honor and love
her husband.

Some, perhaps, may not think that it becomes
a woman to philosophize, any more than it is suitable for her to ride on horseback, or to harangue
in public. But I think that while there are certain employments specialized to each sex, that
there are some common to both man and woman, while
while some belong to a sex only preferentially.
Male avocations are to lead an army, to govern,
and to harangue in public. Female avocations are
to guard the house, to stay at home, to receive
and minister to her husband. Her particular virtues are fortitude, justice and prudence. Both
husband and wife should achieve the virtues of
the body and the soul; for as bodily health is
beneficial to both, so also is health of the
soul. The bodily virtues, however, are health,
strength, vigor of sensation, and beauty. With
respect to the virtues, also, some are peculiarly suitable to men, and some to women. Fortitude
and prudence regard the man more than they do th
the woman; both on account of the bodily habits,
and the soul-power; but temperance peculiarly
belongs to the woman.

It would be well to know the number and quality of the things through which this virtue is
acquirable by women. I think that they are five.
First, temperance comes through the sanctity and
piety of the marriage bed. Second, through body-
ornaments; third, through trips outside the
house. Fourth, through refraining from celebrat-

ing the orgies and mysteries of Cybele. Fifth,
in being cautious and moderate in sacrifices to
the divinities. Of these, however, the greatest and most comprehensive cause of temperance
is undefiledness in the marriage bed; and to have
connexion with none but her husband.

By such lawlwssness she acts unjustly towards
the Gods who preside over nativities, changing
them from genuine to spurious assistants to her
family and kindred. In the second place, she acts
unjustly towards the g ods who preside over Nature, by whom she and all her kindred solemnly
swore that she would lawfully associate with
her husband in the association of life, and the
procreation of children. Third, she injures her
country, in not observing its decrees It is
frivolous and unpardonable, for the sake of pleasure and wayward insolence, to offend in a matter
where the crime is so great that the greatest
punishment, death, is ordained. All such insolent
conduct ends in death. Besides, for this offence
there has been discovered no purifying remedy;
which might turn such guilt into purity beloved
by the divinity, for God is most averse to the
pardoning of this crime. The best indication of
a woman's chastity towards her husband is her
children's resemblance to their father. This suffi-
ces about the marriage-bed.

As to body-ornaments, a woman's garments should
be white and simplem and not superfluous. They will
be so if they are neither transparent nor variegat-
ed, nor woven from silk, inexpensive, and white.
This will prevent excessive ornamentation, luxury,
and superfluity of clothes; and will avoid the imitation of depravity by others. Neither gold nor
emeralds should ornament her body; for they are
very expensive, and exhibit pride and arrogance toward the vulgar. Besides, a city governed
by good laws, and well organized, should adjust all i
all its interests in an equable legislation; which
therefore would expel from the city the jewelers
who make such things.

A woman should, besides, illuminate her face,

not by powder or rouge, but by the natural glow
from the towel, adorning herself with modesty,
rather than by art. Thus she will reflect honor
both on herself and her husband.

As to gadding, women should chiefly go out of
their houses to sacrifice to the municipal tutelary divinity , for the welfare of her husband and
her kindred. Neither should a woman go out from
her house at dawn or dusk, but openly when the forum is full of people; accompanied by one, at most
two servants, to see something, or to shop.

As to sacrifices of the gods, they should be fru
gal, and suited to her ability; she should abstain
from celebration of orgies, and the Cybalean sacre
rites performed at home. For the municipal law for
bids them to women. Moreover, these rites lead to
intoxibation and insanity. A family-mistress, pre
siding over domestic affairs, should be temperate
and undefilod.

CLINIAS

Every virtue is perfected, as was shown in
the beginning, by reason, deliberate choice,
and power. Each of these, however, is by itself
not a part of virtue, but its cause. Such, ther
therefore, as have the intellective and gnostic
part of virte (the theoretic virtues), are called skilful and intelligent; but such as have
its ethical and preparatory parts, are called
useful and equitable. Since, however, man is
naturally adapted to act unjustly from exciting causes, these are three: the love of pleasure of corporeal enjoyments, avarice in the accum-
ulation of wealth, and ambition in surpassing
equals or fellows. Now it is possible to oppose
to these such things as procure fear, shame, or
desire in men; fear through the laws, shame through
the Gods, and desire through the energies of
reason. Hence youth should be taught from the very
first to honor the Gods and the laws. Following
these, every human work, and every kind of human
life, by the participation of sanctity and piety,
will sail prosperously over the sea (of generation).

SEXTUS THE PYTHAGOREAN

SELECT SENTENCES

1 To neglect things of the smallest consequence is not the least thing in human life.

2, The sage and the contemner of wealth most resemble God.

3. Do not investigate the name of God, because you will not find it. For everything called by a name receives its appellation from that which is more worthy than itself, so that it is one person that calls, and another that hears. Who is it, therefore, who has given a name to God? The word "God" is not a name of his, but an indication of what we conceive of him.

4, God is a light incapable of receiving its opposite, (darkness).

5. You have in yourself something similar to God, and therefore use yourself as the temple of God, on account of that which in you resembles God.

6. Honor God above all things, that he may rule over you.

7. Whatever you honor above all things, that which you so honor will have dominion over you. But if you give yourself to the domination of God, you will thus have the dominion over all things.

8. The greatest honor which can be paid to God is to know and imitate him.

9. There is not any thing, indeed, which wholly resembles God; nevertheless the imitation of him as much as possible by an inferior nature is grateful to him.

10. God, indeed, is not in want of anything; but the wise man is in want of God alone. He, therefore, who is in want of but few things, and those necessary, emulates him who is in want of nothing.

11. Endeavor to be great in the estimation of divinity; but among men avoid envy.

12. The sage whose estimation with men was but small while he was living, will be renewed when he is dead.

13. Consider lost all the time in which you do not think of divinity.

14. A good intellect is the choir of divinity.

15. A bad intellect is the choir of evil geniuses.

16. Honor that which is just, on this very account that it is just.

17. You will not be conceiled from divinity when you act unjustly, nor even when you think of acting so.

18. The foundation of piety is continence, but the summit of piety is love to God.

19. Wish that what is expedient and not what is pleasing may happen to you.

20. Such as you wish your neighbor to be to you, such also be to your neighbors.

21. That which God gives you none can take away.

22. Neither do, nor even think, of that which you are unwilling God should know.

23. Before you do anything, think of God, that his light may precede your energies.

24. The soul is illuminated by the recollection of God.

25. The use of animal food is indifferent, but it is more rational to abstain from them.

26. God is not the author of any evil.

27. You should not possess more than the use of the body requires.

28. Possess those things that no one can take away from you.

29. Bear that which is necessary, as it is necessary.

30. Ask of God things such as it is worthy of God to bestow.

31. The reason that is in you is the light of your life.

32. Ask from God those things that you cannot receive from man.

33. Wish that those things which labor ought to precede, may be possessed by you after labor.

34. Be not anxious to please the multitude.

35. It is not proper to despise those things

of which we shall be in want after the dissolution
of the body.

36. Do not ask of divinity that which, when
you have obtained, you cannot perpetually possess.

37. Accustom your soul after (it has conceived all that is great of)divinity, to conceive
something great of itself.

38. Esteem precious nothing which a bad man
can take from you.

39. He is dear to divinity, who considers
those things alone precious, which are esteemed
to be so by divinity.

40. Everything superfluous is hostile.

41. He who loves that which is not expedient,
will not love that which is expedient.

42. The intellect of the sage is always with
divinity.

43. God dwells in the intellect of the wise
man.

44. The wise man is always similar to himself.

45. Every desire is insatiable, and therefore
is always in want.

46. The knowledge and imitation of divinity
are alone sufficient to beatitude.

47. Use lying as poison.

48. Nothing is so peculiar to wisdom as truth.

49. When you preside over men, remember that
divinity presides over you also.

50. Be persuaded that the end of life is to
live conformably to divinity.

51. Depraved affections are the beginnings
of sorrows.

52. An evil disposition is the disease of the
soul; but injustice and impiety is the death
of it.

53. Use all men in a way such   as if, after
God, you were the common curator of all things.

54. He who uses badly mankind, badly uses himself.

55. Wish that you may be able to benefit your
enemies.

56. Endure all things, in order that you may
live conformably to God.

57. By honoring a wise man, you will honor
yourself.

58. In all your actions, keep God before your eyes.

59. You may refuse matrimony, in order to live in incessant presence with God. If, however, you know how to fight, and are willing to, take a wife, and beget children.

60. To live, indeed, is not in our power; but to live rightly is.

61. Be unwilling to entertain accuseations against a man studious of wisdom.

62. If you wish to live successfully, you will have to avoid much, in which you will come out only second-best.

63. Sweet to you should be any cup that quenches thirst.

64. Fly from intoxication as you would from insanity.

65. No good originates from the body.

66. Estimate that you are suffering a great punishment when you obtain the object of corporeal desire; for desire will never be satisfied with the attainments of any such objects.

67. Invoke God as a witness to whatever you do.

68. The bad man does not think that there is a Providence.

69. Assert that your true man is he who in you possesses wisdom.

70. The wise man participates in God.

71. Wherever that which in you is wise resides, there also is your true good.

72. That which is not harmful to the soul does not harm the man.

73. He who unjustly expels from his body a wise man, by his iniquity confers a benefit on his victim; for he thus is liberated from his bonds.

74. Only through soul-ignorance is a man saddened by fear of death.

75. You will not possess intellect till you understand that you have it.

76. Realize that your body is the garment of your soul; and then you will preserve it pure.

77. Impure geniuses let not the impure soul escape them.

78. Not to every man speak of God.

79. There is danger, and no negligeable one, to speak of God even the things that are true.

80. A true assertion about God is an assertion of God.

81. You should not dare to speak of God to the multitude.

him 82. He who does not worship God does not know him.

83. He who is worthy of God is also a god among men.

84. It is better to have nothing, than to possess much, and impart it to no one.

85. He who thinks that there is a God, and that he protects nothing, is no whit better than he who does not believe there is a God.

86. He best honors God who makes his intellect as like God as possible.

87. Who injures none has none to fear.

88. No one who looks down to the earth is wise

89. To lie is to deceive, and be deceived.

90. Recognize what God is, and that in you which recognizes God.

91. It is not death; but a bad life, which destroys the soul.

92. If you knew Him by whom you were made, you would know yourself.

93. It is not possible for a man to live conformably to Divinity, unless he acts modestly, well and justly.

94. Divine wisdom is true science.

95. You should not dare to speak of God to an impure soul.

96. The wise man follows God, and God follows the soul if the wise man.

97. A king rejoices in those he governs, and therefore God rejoices in the wise man. He who governs likewise, is inseparable from those he governs; and therefore God is inseparable from the soul of the wise man, which He defends and governs.

98. The wise man is governed by God, and on this account is blessed.

99. A scientific knowledge of God causes a man to use but few words.

100. To use many wordsi in speaking of God obscures the subject.

101. The man who possesses a knowledge of God will not be very ambitious.

102. The erudite, chaste and wise soul is the prophet of the truth of God.

103. Accustom yourself always to look to the Divinity.

104. A wise intellect is the mirror of God.

(These sentences were preserved by Rufinus,-a Christian writer, who would not have taken the trouble to do so unless indeed their intrinsic worth had been as great as it is.)

SELECT PYTHAGOREAN SENTENCES.

1. From the PROTREPTICS OF IAMBLICHUS

105. As we live through soul, it must be said that by the virtue of this we do live well; just as because we see through the eyes, we see well through their virtues.

106. It must not be thought that gold can be injured by rust, or virtue by baseness.

107 We should betake ourselves to virtue as to an inviolable temple, so that we may not be exposed to any ignoble insolence of soul, with respect to our communion with, and continuance in life.

108. We should confide in virtue as in a chaste wife; but trust to fortune as to an inconstant mistress.

109. It is better that virtue should be received accompanied by poverty, than wealth with violence; and frugality with health, than veracity with disease.

110. An overabundance of food is harmful to the body; but the body is preserved when the soul is disposed in a becoming manner.

111. It is as dangerous to give power to a depraved man, than a sword to a madman.

112. As it is better for a part of the body that contains purulent decay to be burned, than to continue as it is, thus also is it better for a depraved man to die, than to continue to live.

113. The theorems of philosophy are to be enjoyed as much as possible, as if they were ambrosia and nectar. For the resultant pleasure is genuine, incorruptible and divine. They are also capable of producing magnanimity, and though they cannot make us eternal, yet they enable us to obtain a scientific knowledge of eternal natures.

114. If vigor of sensation is, as it is, considered to be desirable, so much more strenuously should we endeavor to obtain prudence; for it is, as it were, the sensitive vigor of the practical

intellect, which we contain. And as through the
former we are not deceived in sensible percep-
tions, so through the latter we avoid false reasonings in practical affairs.

115. We shall prperly venerate Divinity if
we purify our intellect from vice, as from a stain
stain.

116. A temple should, indeed, be adorned with
gifts; but our soul with disciplines.

117. As the lesser mysteries are to be delivered before the greater, thus also discipline:
must precede philosophy.

118. The fruits of the earth, indeed, appear
annually; but the fruits of philosophy ripen at
all seasons.

119. As he who wishes the best fruit must
pay most attention to the land, so must the greatest attention be paid the soul, if it is to prod
duce fruits worthy of its nature.

2. FROM STOBAEUS

120. Do not even think of doing what ought
not to be done.

121. Choose rather to be strong in soul, than
in body.

122. Be sure that laborious things contribute
to virtue, more than do pleasurable things.

123. Every passion of the soul is most hostile
to its salvation.

124. Pythagoras said that it is most difficult simultaneously to walk in many paths of life.

125. Pythagoras said that we must choose the
best life; for custom will make it pleasant. Wealth
is a weak anchor; glory, still weaker; and similarly with the body, dominion, and honor. Which
anchors are strong? Prudence, magnanimity and fortitude; these can be shaken by no tempest. This is
the law of God, that virtue is the only thing
strong; all else is a trifle. (Taylor thinks that
this and the next six sentences are wrongly attributed to Socrates, and are by Democrates or Demophilus).

1266. All theeparts of human life, just as those of a statue, should be beautiful.

127. As a statue stands immovable on its pedestal, so should stand a man on his deliberate choice, if ho is worthy.

128. Incense is for the Gods, but praise to good men.

129. Men unjustly accused of acting unjustly should be should be defended, while those who excel should be praised.

130. It is not the sumptuous adornment of the horse that earns him praise, but the nature of th horse himself; nor is the man worthy merely because he owns great wealth, but he whose soul is generous.

131. When the wise man opens his mouth the beauties of his soul present themselves to view as the statues in a temple (when the gates are opened.

132. Remind yourself that all men assert that wisdom is the greatest good, but that there are few who strenuously endeavor to obtain this greatest good. — Pythagoras.

133. Be sober, and remember to be disposed to believe; for these are the nerves of wisdom. — Epicharmus.

134. It is better to live lying on the grass, confiding in divinity and yourself, than to lie on a golden bed with perturbation.

135. You will not be in want of anything, which it is in the power of Fortune to give er take away.— Pythagoras.

136. Despise all those things which you will not want when liberated from the body; and exercising yourself in those things of which you will be in want, when liberated from the body, be sure to invoke the Gods to become your helpers. — Pythagoras.

137. It is as impossible to conceal fire in a garment, as a base deviation from rectitude in time.— Demophilus, rather than Socrates.

138. Wind increases fire, but costom, love. Ibidem.

139. Only those are dear to divinity, who are hostile to injustice. es Democritus or Demophilus.

140. Bodily necessities are easily procured by anybody; without labor or molestation; but those things whose attainment demands effort and trouble, are objects of desire not to the body, but to deprave praved opinion. Aristoxenus the Pythagorean.

141. Thus spoke Pythagoras of desire: This passion is various, laborious and very multiform. Of des desires, however, some are acquired and artificial, while others are inborn. Desire is a certain tendency and impulse of the soul, and an appetite of fulness, or presence of sense, or of an emptiness and absence of it, and of non-perception. Thre three best known kinds of depraved desire are the improper, the unproportiona te, and the unseasonable. For desire is either immediately indecorous, troublesome or illiberal; or if not absolutely so, it is improperly vehement and persistent. Or, in the third place, it is impelled at an improper time, or towards improper objects. — Aristoxenus.

142. Pythagoras said, Endeavor not to conceal your errors by words, but to remedy them by reproofs.

143. Pythagoras said, It is not so difficult to err, as not to reprove him who errs.

144. As a bodily disease cannot be healed, if it is concealed or praised, thus also can neither a remedy be applied to a diseased soul, which is badly guarded and protected. — Pythagoras.

145. The grace of freedom of speech, like beauty in season, is productive of greater delight.

146. To have a blunt sword is as improper as to use ineffectual freedom of speech.

147. As little could you deprive the world of the sun, as freedom of speech from erudition.

148. As one who is clothed with a cheap robe may have a good body-habit, thus also may he whose life is poor possess freedom of speech.

149. Pythagoras said, Prefer those that reprove, to those that flatter; but avoid flatterers as much as enemies.

150. The life of the avaricious resembles a funeral banquet. For though it has all desirable elements no one rejoices.

151. Pythagoras said: Acquire continence as the greatest strength and wealth.

152. "Not frquently man from man," is one of the exhortations of Pythagoras; by which obscurely he signifies that it is not proper frequently to engage in sexual connections.

153. Pythagoras said: A slave to his passions cannot possibly be free.

154. Pythagoras said that intoxication is the preparation for insanity.

155. On being asked how a wine-lover mightbe cured of intoxication Pythagoras said, "If he frequently considers what were his actions during intoxication.

156. Pythagoras said that unless you had something better than silence to say, you had better keep silence.

157. Pythagoras said, that rather than utter an idle word you had better throw a stone in vain.

158. Pythagoras said, "Say not few things in many words, but much in few words.

159. Epicharmus said, " To men genius is a divinity either good or evil.

160. On being asked how a man ought to behave towards his country when it had acted unjustly towards him, Pythagoras said, "As to a mother."

161. Traveling teaches a man frugality, and self-sufficiency. The sweetest remedies for hunger and weariness are bread made of milk and flour, on a bed of grass(Democritus, probably Democrates or Demophilus; also the next one).

162. Every land is equally suitable as a residence for the wise man; the worthy soul's fatherland is the whole world.

163. Pythagoras said that into cities entered first, luxury; then being glutted, then lascivious insolence, and last destruction.

164. Pythagoras said that that was the best city which contained the worthiest men.

165. Pythagoras added to Demophilus's maxim

THAT "You should do these things that you judge
to be beautiful, though in doing them you should
lack renown; for the rabble is a bad judge of a
good thing," the words, "Therefore despise the
reprehension of those whose praise you despise."

166. Pythagoras said that Those who do not punish bad men, are really wishing that good men
be injured.

167. Pythagoras said, Not without a bridle
can a horse be gayarned, and no less riches without prudence.

168. The prosperous man who is vain is no better than the driver of a race on a slippery road
(Socrates? Probably Democrates, or Demophilus).

169. There is no gate of wealth so secure
but that may open to the opportunity of Fortuen
(Democritus? Probably Democrates or Demophilus).

170. The unrestrained grief of a torpid soul
may be expelled by reasoning. (Democrates, not
Democritus).

171. Poverty should be born with equanimity
by a wise man. (Same).

172. Pythagoras; Spare your life, lest you consume it with sorrow and care.

173. Phavorinus, in speaking of Old Age, said,
Nor will I be silent as to this particular, that
both to plato and Pythagoras it appeared that old
age was not to be considered with reference to an
egress from the present life, but to the beginning
of a blessed one.

§. From CLEMENT OF ALEXANDRIA, Strom.3:413.

174. Philolaus said that the ancient theologians
and priests testified that the soul, is united
to the body by a caertain punishment, and that it
is buried in this body as a sepulchre.

175. Pythagoras said that "Whatever we see
when awake is death, and when asleep is a dream.

HIEROCLES,

ETHICAL FRAGMENTS (preserved by STOBAEUS).

(His Commentary on the Golden Verses is wordy
and common-place, and therefore is here omitted)

I.

CONDUCT TOWARDS THE GODS.

Concerning the Gods we should assume that they
are immutable, and do not change their decrees,
from the very beginning they never vary their
conceptions of propriety. The immutability and
firmness of the virtues we know, and reason
suggests that it must transcendently obtain wit
with the Gods, and be the element which to their
conception imparts a never-failing stability.
Evidently no punishment which divinity thinks
proper to inflict is likely to be remitted. For
if the Gods changed their decisions, and omitted
to punish someone whom they had designed to punish, the world could be neither beautifully nor
justly governed; nor can we assign any probable
reason for repentance (on their part). Rashly,
indeed, and without any reason, have poets written words such as the following:

"Men bend the Gods, by incense and libation,

By gentle vows, and sacrifice and prayer,
When they transgress, and stray from what is
right!" (Homer, Iliad, ix: 495-7).
And: "Flexible are e'en the Gods themselves!" (493)

Nor is this the only such expression in poetry.
Nor must we omit to observe, that though the
Gods are not the causes of evil, yet they connec'
certain persons with things of this kind, and su
round those who deserve to be afflicted with cor
poreal and external hindrances; not through any
malignity, or because they think it advisable

THAT MEN SHOULD STRUGGLE WITH difficulties, but for the sake of punishment. For as in general pestilence and drought, rain-storms, earth-quakes and the like, are indeed for the most part produced by natural causes, and yet are sometimes caused by the Gods, when the times are such that the multitude's iniquity needs to be punished publicly, and in common, likewise in particular the Gods sometimes afflict an individual with corporeal and external difficulties, in order to punish him, and convert others to what is right.

The belief that the Gods are never the cause of any evil, it seems to me, contributes greatly to proper conduct towards the Gods. For evils proceed from vice alone, while the Gods are of themselves the causes of good, and of any advantage; though in the meantime we slight their beneficence, and surround ourselves with voluntary evils. That is why I agree with the poet who says,
"That mortals blame the Gods,....
as if they were the causes of their evils!
"Though not from fate,
"But for their crimes they suffer woe!"
(Homer Odyssey, i.32-34).
. Many arguments prove that God is never in anynway way the cause of evil; but will suffice to read (in the first book of the Republic) the words of Plato, "that as it not the nature of heat to refrigerate, so the beneficent cannot harm; but the contrary." Moreover, God being good, and from the beginning replete with every virtue, cannot harm, not cause evil to any one; on the contrary, imparting good to all willing to receive it; bestowing on us also such indifferent things as flow from nature, and which result in accordance with nature. But there is only one cause of evil.

II.

PROPER CONDUCT TOWARDS OUR COUNTRY.

After speaking of the Gods, it is most reasonable, in the second place, to show how we should conduct ourselves towards our country. For God is my witness that our country is a sort of secondary divinity, and our first and greatest parent. That is why its name is, for good reason, patris, derived from pater, a father; but taking a feminine termination, to be as it were a mixture of father and mother. This also explains that our country should be honored equally with our parents: preferring it to either of them separately, and no not even to it preferring both our parents; preferring it besides to our wife, children and friends; and in short to all things, under the Gods.

He who would esteem one finger more than five would be considered stupid; inasmuch as it is reas reasonable to prefer five to one; the former despising the most desirable, while the latter, among the five preserves also the one finger. Likewise, he who prefers to save himself rather than his country, in addition to acting unlawfully, desires an impossibility. On the contrary, he who to himself prefers his country is dear to divinity, and reasons properly and irrefutably. Moreover it has been observed that though someone should not be a member of an organized society, remaining apart therefrom, yet is it proper that he should prefer the safety of society to his own; for the city's destruction would demonstrate that on its existence depended that of the individual citizen, just as the amputation of the hand involves the destruction of the finger, as an integral part. We may therefore draw the general conclusion that general utility cannot be separated from private welfare, both at bottom being identical. For whatever is beneficial to the whole country

is common to every single part, inasmuch as without the parts the whole is nothing. Vice versa, whatever redounds to the benefit of the citizen extends also to the city; the nature of which is to extend benefits to the citizen. For example, whatever id beneficial to a dancer, must, in so far as he is a dancer, be so also to the whole choric ballet. Applying this reasoning to the discursive power of the soul, it will shed light on every particular duty, and we shall never omit to perform whatever may by us be due to our country.

That is the reason why a man who proposes to act honorably by his country should from his soul remove every passion and disease. The laws of his country should, by a citizen, be observed as (precepts of) a secondary divinity, conforming himself entirely to their mandates. He who endeavors to transgress or make any innovation in these laws should be opposed in every way, and be prevented therefrom in every possible way. By no means beneficial to a city, is contempt of existing laws, and preference for the new. Incurable innovators, therefore, should be restrained from giving their votes, and making precipitate innovations. I therefore comment the Locrian legislator Zaleucus, who ordained that he who intended to introduce a new law should do it with a rope around his neck, in order that he might be immediately strangled unless he succeeded in changing the ancient constitution of the state, to the very great advantage of the community.

But customs which are truly those of the country, and which, perhaps, are more ancient than the laws themselves, are, no less than the laws, to be preserved. However, the customs of the present, which are but of yesterday, and which have been everywhere introduced only so very recently, are not to be dignified as the institutes of our ancestors, and perhaps they are not even to be considered customs. Moreover, because custom is an unwritten law, it has as sanction the authority of a very good legislator, namely, common consent c of all that use it; and perhaps on this account its authority is next to that of justice itself.

III.

PROPER CONDUCT TOWARDS THE PARENTS.

After considering the Gods and our parents, what person deserves to be mentioned more than, or prior to our parents? That is why we turn towards them. No mistake, therefore, will be made by him who says that they are as it were secondary or terrestrial divinities, since, on account of their proximity, they should, in a certain not blasphemous sense, be by us more honored than the Gods themselves. To begin with, the only gra gratitude worthy of the name is a perpetual and unremitting promptness to repay the benefits received from them; since, though we do our very utmost, this would yet fall short of what they deserve. Moreover, we might also say that in one sense our deeds are to be counted as theirs, because we who perform them were once produced by them. If, for instance, the works of Phidias and other artists should themselves produce other works of art, we should not hesitate to attribute these latter deeds also to the original artists; that is why we may justly say that our performances are the deeds of our parents, through whom we originally derived our existence.

In order that we may the more easily apprehend the duties we owe them, we should keep in mind the underlying principle, that our parents should by us be considered as the images of the Gods; and, so help me heaven, images of gods domestic, who are our benefactors, our relatives, our creditors, our lords, and our most stable friends. They are indeed most stable images of the gods, possessing a likeness to them which no artist could possibly surpass. They are the guardian divinities of the home, and live with us; they are our greatest benefactors, endowing us with benefits of the greatest consequence, and indeed bestowing on us not only all we possess, but also such things as they wish to give us, and for which they themselves pray. Further they are our nearest kindred, and the causes of our alliance with others. They are also creditors of things of the most honorable nature, and repay themselves only by taking what we shall be benfitted by returning. For to a child what benefit can be so great as piety and gratitude to his parents? Most justly, too, are they our lords, for of what can we be the possession in a dgree greater than of those through whom we exist? Moreover,

they are perpetual and spontaneous friends and
auxiliaries, affording us assistance at all times
and in every circumstance. Since, besides, the
name of parent is the most excellent of names,
which we apply even to the divinities, we may
add something further to this conception: namely,
that children should be persuaded that they dwell
in their father's house, as if they were ministers
and priests in a temple, appointed and consecrated
for this purpose by nature herself, who entrusted
to their care a reverential attention to their
parents. If we are willing to carry out the dict-
ates of reason, we shall readily attend to both
kinds of affective regard, that regarding the
body and the soul. Yet reason will show us that
to the body, is to be paid less regard than to
the soul, althoug we shall not neglect the former
very necessary duties. For our parents, therefore,
we should obtain liberal food, and such as is ada-
apted to the weakness of old age; besides this, a
bed, sleep, massage, a bath, and proper garments;
in short, the necessaries of the body, that they
may at no time experience the want of any of these;
by this imitating their care for the nurture of
ourselves, when we were infants. Our attention
to them should partake of the prophetic nature,
whereby we may discover what special bodily
necessity they may be longing, without expressing
it to us. Respecting us, indeed, thet divined many
things, when our desires bould be expressed by
no more than inarticulate and distressful cries,
unable to express the objects of our wants clealy.
By the benefits they formerly conferred upon us
our parents became to us the preceptors of what
we ought to bestow upon them.

With respect to our parents' souls, we should,
in the first place, procure for them diversion,
which will be obtained especially if we associate
with them by night and day, taking walks, being
massaged, and living by their side, unless something necessary interferes. For just as those
who are undertaking a long journey desire the
presence of their families and friends to see
them off, as if accompanying a solemn procession,

So also parents, verging on the grave, enjoy most
of all the sedulous and unremitting attention of
their children. Moreover, should our parents at
any time, as happens often, especially with those
whose education was deficient, their conduct show
be reprehensible, they should indeed be correcte
but not as we are accustomed to do with our inferiors or equals, but as it were with suggest-
iveness; not as if they had erred through ignorance, but as if they had committed an oversight
through inattention, as if they would not have
erred, had they considered the matter. For reproof, especially if personal, is to the old ve
bitter. That is why their oversights should be
supplemented by mild exhortation, as by an eleg
ant artifice.

Children, besides, rejoice their parents by
performing for them servile offices such as wash
ing their feet, making their bed, or ministering
to their wants. These necessary servile attentia
are all the more precious when performed by the
dear hands of their children, accepting their
ministrations. Parents will be especially gratified when their children publicly show their
honir to those whom they love and very much este

That is why children should affectionately
love their parents' kindred, and pay them proper
attention, as also to their parents' friends and
acquaintances. These general principles will aid
us to deduce many other smaller filial duties,
which are neither unimportant nor accidental. Fc
since our parents are gratified by the attention
we pay to those they love, it will be evident tha
as we are in a most eminent degree beloved by
our parents, we shall surely much please them by
paying a proper attention to ourselves.

IV.

ON FRATERNAL LOVE

The first admonition, therefore, is very cl
and convincing, and obligatory generally, bein

SANE and self-evident. Here it is: Act by every one, in the same manner as if you supposed yourself to be him, and him to be you. A servant will be well treated by one who considers how he would like to be treated by him, if he was the master, and himself the servant. The same principle might be applied between parents and children, and vice versa; and in short, between all men. This principle, however, is peculiarly adapted to the mutual relation of brothers; since no other preliminary considerations are necessary, in the matter of conduct towards one's brother, than promptly to assume that equable mutual relation. This therefore is the first precept, to act towards one's brother in the same manner in which he would think it proper for his brother to act towards him.

But some one will say, I do not transgress propriety, and am equitable; but my brother's manners are rough and brusque. This is not right; for, in the first place, he may not be speaking the truth; as excessive vanity might lead a man to extol and magnify his own manners, and diminish and vilify what pertains to others. It frequently happens, indeed, that men of inferior worth prefer themselves to others who are far more excellent characters. Second, though the brother should indeed be of the rough character mentioned above, the course to take would be to prove oneself the better character, by vanquishing his boorishness by your dutifulness. Those who conduct themselves worthily towards moderate, benignant men are entitled to no great thanks; but to transform to graciousness the stupid vulgar man, he deserves the greatest applause.

It must not be thought impossible for exhortation to take marked effect; for in men of the most impossible manners there are possibilities of improvement, and of love and honor for their benefactors. Not even animals, and such as naturally are the most hostile to our race, who are captured by violence and dragged off in chains, and confin in cages, — are not beyond being tamed by appropriate treatment, and daily food. Will not then th

man who is a brother, or even the first man you
meet, who deserves attention far greater than a
brute, be rendered gentle by proper treatment,
even though he should never entirely lose his
boorishness? In our behavior, therefore, towards
every man, and in much greater degree towards a
brother, we should imitate Socrates who, to a pe.
on who cried out against him, "May I die, unless
am revenged on you," answered, "May I die, if I do
not make you my friend! So much then for externa
fraternal relations.

Further, a man should consider that in a certe',
sense his brothers are part of him, just as my e
are part of me; also my legs, my hands, and othe
parts of me. For the relation of brothers to a
family social organism (are the same as members
a body). If then the eyes and the hands should
ceive a particular soul and intellect, they woul
because of the above mentioned communion, and b
cause they could not well perform their proper o.
fices without the presence of the other member.
watch over the interests of the other members w
the interest of a guardian genius. So also, we
who are men, and who acknowledge that we have a
soul, should, towards our brothers, omit no prop
offices. Indeed, more naturally adapted for muti
assistance than parts of the body, are brothers.
The eyes, being mutually adjusted, do see what i
before them, and one hand cooperates with the ot
er; but the mutual adaptation of brothers is far
more various. For they accomplish things which
are mutually prefitable, though at the greatest
intervening distance; and they will greatly bene
it each other though their mutual difference be
immeasurable. In short, it must be considered, t
our life resembles nothing so much as a prolonged
conflict, which arises partly from the natural
strife in the nature of things, and partly throu
the sudden unexpected blows of fortune; but most
of all through vice itself, which abstains neith
from violence, fraud, or evil stratagems. Hence
nature, as being not ignorant of the purpose for
which she generated us, produced each of us as i
were accompanied by an auxiliary.

No one, therefore, is alone, nor does he derive his origin from an oak or a rock, but from parents, in conjunction with brothers, relatives, and other intimates. Here reason for us performs a great work, conciliating to us strangers, who are no relatives of ours, furnishing us with many assistants. That is the very reason why we naturally endeavor to allure and make every one our friend. How insane a thing it therefore is to wish to be united to those who naturally have nothing suitable to procure our love, and become as familiar as possible with them, voluntarily; and yet neglect those willing helpers and associates supplied by nature herself, who are called brothers!

V.

ON MARRIAGE

The discussion of marriage is most necessary, as the whole of our race is naturlaay social; and the most fundamental social association is that effected by marraige. Without a household, there could exist no cities; and households of the unmarried are most imperfect, while on the contrary those of the married are most complete. That is why, in our treatise on Families, we have shown that the married state is to be preferred by the sage; while a single life is not to be chosen except under peculiar circumstances.(Pyhtagoras and Socrates were married, while Plato, Plotinus and Proclus were not). Therefore, inasmuch as we should imitate the man of intellect so far as possible, and as for him marriage is preferable, it is evident it will be so also for us, except if hindered by some exceptional circumstance. This is the first reason for marriage.

Entirely apart from the model of the sage, Nature herself seems to incite us thereto. Not only did she make us gregarious, but adapted us to sexual intercourse, and proposed the procreation of children and stability of life as the one and universal work of wedlock. Now Nature justly teaches us that a choice of such things as are fit should be made so as to accord with what she has procured

for us. Every animal, therefore, lives in conform-
ity to its natural constitution, and so also every
plant in harmony with its laws of life. But there
obtains this difference: that the latter do not employ any reasoning or calculation, in the selection
of the things on which they lay hold, using alone
nature, without participatingnin soul. Animals are
drawn to investigate what may be proper for them
by imaginations and desires. To us, however, Nature
gave reason, to survey everything else, and, together with all things, nay, prior to all things,
to direct its attention to Nature itself, so as
to tend towards her, as a glorious aim, in an orderly manner, that by choosing everything consonant
with her, we might live in a becoming manner. Following this line of argument, he will not err in
saying that a family without wedlock is imperfect;
for (nature) does not conceive of the governor
without the governed, nor the governed without a
governor. Nature therefore seems to me to shame
those who are averse to marriage.

In the next place, marriage is beneficial. First,
because it produces a truly divine fruit, the
procreation of children, who are, as partaking of
our nature, to assist us in all our undertakings,
while our strength is yet undiminished; and when
we shall be worn out, oppredsed with old age, they
will be our assistants. In prosperity they will
be the associates of our joy, and in adversity,
the sympathetic diminishers of our sorrows.

Marriage is beneficial not only because of
procreation of children, but for the association
of a wife. When we are wearied with our labors
outside of the home, she receives us with offici-
ous kindness, and refreshes us by her solicitous
attentions. Next, she induces a forgetfulness of
molestations outside of the house. The annoyances
in the forum, the gymnasium, or the country, and
in short all the vicissitudes of our intercourse
with friends and acquaintances, do not disturb us
so obviously, being obscured by our necessary occupations; but when released from these, we return
home, and our mind has time to reflect, then, avail-
ing themselves of this opportunity these cares and

anxieties rush in upon us, to torment us, at the
very moment when life seems cheerless and lonely.
Then comes the wife as a great solace, and by making some inquiry about external affairs, or by
referring to, and together considering some domestic problem, she, by her sincere vivacity inspires
him with pleasure and delight. It is needless to
enumerate all the help a wife can be in festivals,
when sacrificing victims; or during her husband's
journeys, she can keep the household running
smoothly, and direct at times of urgency; in
managing the domestics, and in nursing her husband
when sick.

Summarizing, in order to pass through life
properly, all men need two things: the aid of rela
atives, and kindly sympathy. But nothing can be
more sympathetic than a wife; nor anything more
kindred, than childre. Both of these are afforded by marriage; how therefore could we find anyt
thing more beneficial?

Also beautiful is a married life, it seems to
me. What relation can be more ornamental to a family, than that between husband and wife? Not sumptu
uous edifices, not walls covered with marble plas-
ter, not piazzas adorned with stones, which are indeed admired by those ignorant of true goods; not
paintings and arched myrtle walks, nor anything
else which is the subject of astonishment to the
stupid, is the ornament of a family. The beauty of
a household consists in the conjunction of man and
wife, united to each other by destiny, and conseo-
rated to the gods presiding over nuptial births, and
houses, and who harmonize, and use all things in
common for their bodies, or even their very souls;
who likewise exercise a becoming authority over
their house and servants; who are properly solicitous
about the education of their children; and to the
necessaries of life pay an attention which is neither excessive or negligent, but moderate and appropriate. For, as the most admirable Homer says, what
can be better and more excellent,
"Than when at home the husband and wife
"Live in entire unanimity!" (Odyssey, 7:183).

That is the reason why I have frequently wondered at those whoc conceive that life in common
with a woman must be burdensome and grievous. Though
to them she appears to be a burden and molestation,
she is not so; on the contrary, she is something
light and easy to be borne, or rather, she possesses the power of charming away from her husband
things burdensome and grievous. No trouble so great
is there which cannot easily be borne by a husband
and wife who harmonize, and arenwilling to endure t
it in common. But what is truly burdensome and unbearable is imprudence, for through it things naturally light, and among others a wife, become heavy.

To many, indeed, marriage is intolerable, in re
ality not from itself, or because such an association as this with a woman is naturally insufferable,
but when we marry the wrong person, and, in additi
ion to this, are ourselves entirely ignorant of
life, and unprepared to take a wife in a way such
as a free-born woman ought to be taken, than indeed
it happens that this association with her becomes
difficult and intolerable. Vulgar people do marry
in this way; taking a wife neither for the procreation of children, nor for harmonieus association;
being attracted to the union by the magnitude of
the dower, or through physical attractiveness, or
the like; and by following these bad counsellors,
they pay no attention to the bride's disposition
and manners, celebrating nuptials to their own destruction, and with crowned doors introduce to themselves ####### instead of a wife, a tyrant, whom
they cannot resist, and with whom they are unable
to contend for chief authority.

Evidently, therefore, marriage becomes burdensome and intolerable to many, not through itself,
but through these causes. But it is not wise to
blame things which are not harmful, nor to make
our own deficient use of these things the cause of
our complaint against them. Most absurd, besides,
is it feverishly to seek the auxiliaries of friendship, and achieve certain friends and associates,
to aid and defend us in the vicissitudes of life,
without seeking and endeavoring to obtain the relief, defence and assistance afforded us by Nature,
the gods, and the laws, through a wife and childrn.

As to a numerous iffspring, it is generally
suitable to nature and marriage that all, or the
majority of the offspring be nurtured. Many dis-
sent from this, for a not very beautiful reason,
avariciousness, and the fear of poverty as the
greatest evil. To begin with, in procreating
children, we are not only begetting assistants,
nurses for our old age, and associates in every
vicissitude of life; -- we do not however beget
them for ourselves alone, but in many ways also
for our parents. To them our procreation of
children is gratifying; because, if we should
suffer anything calamitous prior to their de-
cease, we shall, instead of ourselves, leave our
children as the support of their old age. Then
for a grandfather it is a beautiful thing to be
conducted by the hands of his grandchildren, and
by them to be considered as worthy of every attention. Hence, in the first place, we shall gratify our own parents by paying attention to the
procreation of children. In the next, we shall be
he cooperating with the ardent wishes and fervent
prayers of those who begot us. They were solicitous about our birth from the first, therethrough
looking for an extended succession of themselves,
that they should leave behind them children of
children, therefore paying attention to our marriage, procreation, and nurture. Hence, by mar-
rying and begetting children we shall be, as it
were, fulfilling a part of their prayers; while,
acting contrariwise, we shall be destroying the
object of their deliberate choice.

Moreover, it would seem that every one who
voluntarily, and without some prohibiting circum
stance avoids marriage and the procreation of
children, accuses his parents of madness, as
having engaged in wedlock without the right conception of things. Here we see an unavoidable con-
tradiction. How could that man live without dissension, who finds a pleasure in living, and wil-
lingly continues in life, as one who was properly
brought into existence bynhis parents, and yet
conceives that for him procreation of offspring
is something to be rejected?

We must remember that we beget children not
only for our own sake, but, as we have already
stated, for our parents; but further also for
the sake of our friends and kindred. It is gratifying to see children which are our offspring
on account of human kindness, relatives, and security. Like ships which, though greatly agitated
by the waves, are firmly secured by many anchors,
so do those who have children, or whose friends
or relatives have them, ride at anchor in port,
in absolute security. For this reason, then, will
a man who is a lover of his kindred, and associates
earnestly desire to marry and beget children.

Our country also loudly calls upon us to do
so. For after all we do not beget children so much
for ourselves, as for our country, procuring a
race that may follow us, and supplying the community with successors to ourselves. Hence the priest
should realize that to the city he owes priests;
the ruler, that he owes rulers; the orator, that
he owes orators; and in short, the citizen, that
he owes citizens. Sa it is gratifying to a choric
ballet that those who compose it should continue
perennially; and as an army looks to the continuance of its soldiers, so the perpetuation of its
citizens is a matter of concern to a city. A city
would not need succession were it only a temporary
grouping, of duration commensurate with the life-
time of any one man; but as it extends to many
generations, and if it invokes a fortunate genius
may endure for many ages, it is evidently necessary to rirect its attention not only to its present, but also to its future, not despising our
natal soil, nor leaving it desolate, but establishing it in good hopes for our posterity.

VI.

CONDUCT TOWARDS OUR RELATIVES.

Duties to relatives depend on duties to our
immediate families, the arguments for which apply
also to the former. Each of us is, indeed, as it
were circumscribed by many circles, larger and
smaller, comprehending and comprehended, according
to various mutual circumstances.

The first and nearest circle is that which every one describes about the centre of his own mind, wherein is comprehended the body, and all its interests; this is the smallest circle, nearly touching the centre itself. The second and further circle which comprehends the first, is that which includes parents, brethren, wife, and children. The third greater circle is the one containing uncles, aunts, grandfathers, and grandmothers, and the children of brothers and sisters. Beyond this is the circle containing the remaining relatives. Next to this is the circle containing the common people, then that which comprehends our tribe, than that of all the citizens; then follow two further circles: that of the neighboring suburbs, and those of the province. The outermost and greatest circle is that which comprehends the whole human race(as repeated Pope, in his Essay on Man).

In view of this, he who strives to conduct himself properly in each of these connections should, in a certain respect, gather together the circles into one centre, and always endeavor to transfer himself from the comprehending circles to the several particulars which they comprehend.

The lover of his kindred, therefore, should conduct himself in a becoming manner towards his parents and brothers; also, according to the same analogy, towards the more elderly of his relatives of both sexes, such as grandfathers, uncles and aunts; towards those of the same age as himself, as his cousins; and towards his juniors, as the children of his cousins. This summarizes his conduct towards his kindred, having already shown how he should act towards himself, his parents and brothers; and besides these, towards wife and children. To which must be added that those who belong to the third circle should be honored similarly to these; and again, kindred similarly to those that belong to the third circle. For benevolence must somehow fade away from those who are more distant from us by blood; though at the same time we should endeavor to effect a mutual assimilation. This distance will moderate if through the diligent attention which we pay to them we shorten the bond connecting us with each. Such then are the most comprehensive duties towards our kindred.

It might be well to say a word about the general names of kindred, such as the calling of cousins, uncles and aunts by the names of brothers, fathers and mothers; while of the other kindred, to call some uncles, others the children of brothers and sisters, and others cousins, according to the difference in age, for the sake of the emotional extension derivable from names. Such name-extension will manifest our sedulous attention to these relatives, and at the same time will incite, and extend us in a greater degree to the contraction of the above circles.

We should however emember the distinction between parents that we madenabove. Comparing parents, we said that to mother was due more love, vut to the father more honor. Similarly, we should show more love to those connected with us by a maternal alliance, but more honor to those connected with us by an alliance that is paternal.

VII.

ON ECONOMICS

To begin with, we must mention the kind of labor which preserves the union if the family. To the hus-husband are usually assigned rural, forensic and political activities; while to the mother belong spinning of wool, making of bread, cooking, and in short, everything of a domestic nature. Nevertheless, neither should be entirely exempt from the labors of the other. For sometimes it will be proper, when the wife is in the country, that she should superintend the laborers, and act as major-demo; and that the husband should sometimes attend to domestic affairs, inquiringnabout, and inspecting what is doing in the house. This joint participation of necessary cares will more firmly unite their m utual association.

We should not fail to mention the manual operations, which are associated with the spheres of
occupations. Why should the man meddle with agricultural labors? This is generally admitted; and
though men of the present day spend much time in
idleness and luxury, yet it is rare to find any
unwilling to engage in the labor of sowing and
planting, and other agricultural pursuits. Much
less persuasive perhaps, will be the arguments
which invite the man to engage in those other occupations that belong to the woman. For such men
as pay great attention to neatness and cleanliness
will not conceive wool-spinning to be their business; since, for the most part vile, diminutive
men, delicate and effeminate apply themselves to
the elaboration of wool, through an emulation of
feminine softness. But it does not become a man,
who is manly, to apply himself to things of this
kind; so that perhaps neither shall I advise such
employments to those who have not unmistakably
demonstrated their modesty and virility. What
therefore should hinder the man from sharing in
the labors pertaining to a woman, whose past life
has been such as to free him from all suspicion
of absurd and effeminate conduct? For is it not
thought that more domestic labors pertain to man
than to women in other fields? For they are more
laborious, and require corporeal strength, such as t
te grind, to knead meal, to cut wood, to draw water
from a well, to carry large vessels from one place
to another, to shake coverlets and carpets, and
such like. It will be quite proper for men to
engage in such occupations.

But it would be well if the legitimate work of
a woman be enlarged in other directions, so that
she may not only engage with her maid-servants in
the spinning of wool, but may also apply herself
to other more virile occupations. It seems to me
that bread-making, drawing water from a well, the
lighting of fires, the making of beds, and such
like are labors suited to a free-bo rn woman.

But to her husband a wife will seem much more
beautiful, especially if she is young, and not yet
worn out by the bearing of children, if she becomes

his associate in the gathering of grapes, and
collecting the olives; and if he is verging toward
olf age, she will render herself more plreasing to
him by sharing with him the labor ef sowing and
plowing, and while he is digging er planting, extending to him the instrumentsnhe needs for his wor.
For when bybthe husband and wife a family is gover.
ernrd thus, in respect to necessary labors, it see.
seens to me that it will be conducted in the best
possible manner.

TIMAEUS LOCRIUS,
The Teacher of Plato, on

THE SOUL AND THE WORLD

1. MIND, NECESSITY, FORM & MATTER

Timaeus the Locrian asserted this: — that of
all the things in the Universe, there are two
causes, (one) Mind, (the cause) of things existing
according to reason; (the other) Nexessity, (the
cause) of things (existing) by (some) force, according to the power of the bodies; and that the
former of these is the nature of the good, and is
called god, and the principle of things that are
best; bur what accessory causes follow, are referred
to Necessity. As regards the things in the Universe,
there are Form, Matter, and the Perceptible; which
is, as it were, a resultance of the two others; and
that Form is unproduced, and unmoved, and stationary,
and of the nature of the same, and perceptible by
the mind, and a pattern of such things produced, as
exist by a staye of change; for that some soh thing
as this is Form, spoken of and conceived to be;

Matter, however, is a mould, and a mother and a
nurse, and procreative of the third kind of being;
for receiving upon itself the resemb,ances, and as
it were remoulding them, it perfects these produc4
tions, He asserted moreover that matter, though
eternal, is not unmoved; and though of itself it
is formless and shapeless, yet it receives every
kind of form; and that what is around bodies, is
divisible, and partakes of the nature of the dif$
ferent; and that matter is called by the twin names
of Placeand Space.

These two principles, then, are opposite to each
other; of which Form relates to a male power, and a
a father; while matter relates to a female, and a
mother. Being three, they are recognizable by three
marks: Form, by mind, according to knowledge; Matter by a spurious kind of reasoning, because of its
not being mentally perceived directly, but by analogy;
and their productions by sensation and opinion.

2. VREATION OF THE WORLD

Before the heavens, then, there existed, throug
reason, Form and Matter, and the God who develops
the best. But since the older surpasses the younge
and the ordered surpasses the orderless, the deity
being good, on seeing that Matter receives Form, a
and is altered in every way, but without order, fe
the necessity of organising it, altering the undef
ined to the defined, so that the differences between bodies might be similarly related, not receiving various turns at hap-hazard. He therefore
made this world out of the whole of Matter, laying
it down as a limitnto the nature of being, throug
its containing in itself all the rest of things,
being one, only-begotten, perfect, endued with
soul and reason, -- for these qualities are superior to the soulless and the irrational, -- and
of a sphere-like body; for this is more pefect
than the rest of forms.

Desirous then of making a very good production
he made it a deity, created and never to be destroyed by any cause other than the God, who had
put it in order, if indeed he should ever wish to
dissolve it. But on the part of the good there is
no rushing forward to the destruction of a very
beautiful production. Such therefore being the
world, it continues withoutcorruption and destruc
tion, being blessed. It is the best af things cre
ed; since it has been produced by the best cause,
that looks net to patterns made by hand, but to
Form in the abstract, and to Existense, perceived
bynthe minds to which the created thing, having
been carefully adjusted, has become the most beau
tiful, and to be not wrongly undertaken. It is ev
perfect according to the things perceived by sens
bec ause the pattern perceived by mind sentains i
itself all the living things perceived by mind; a
he left out of itself nothing, as being thelimit
the things perceived by mind, as this world is of
thoseperceived by sense.

As being solid, and perceptible by touch and
sight, it has a share of earth and fire, and of

the things between them, air and water; and it is
composed of bodies all perfect, which are in it as
wholes, so that no part might ever be left out of
-it, in order that the body of the Universe might
be altogether self-sufficient, uninjured by corruption without or within; for apart from these there
is nothing else, for the things combined according
to the best proportions and with equal powers, neither rule over, nor are ruled by each other in turn,
so that some receive an increase, others a decrease,
remaining indissolubly united according to the very
best proportions.

3. Proportions of the World-combination.

For whenever there are any three terms, with
mutually equal intervals, that are proportionate,
webthen perceive that, after the manner of an
extended string, the middle is to the first, as
is the third to it; and this holds true inversely
and alternately, interchanging places and or der;
so that it is impossible to arrange them numerically without producing an equivalence of results.
Likewise the world's shape and movement are
well arranged; the shape is a sphere self-similar
on all sides, able to contain all shapes that
are similar; the movement endlessly exhibits
the change dependent on a circle. Now as the sphere
sphere is on every side equidistant from the centre, it is able to retain its poise whether in
movement or at rest; neither leaving its poise,
nor assuming another. Its external appearance
being exactly smooth, it needs no mortal organs
such as are fitted to, and present in all other
living beings, because of their wants. The world-
souls element of divinity radiates out from the
centre, entirely penetrating the whole world, forming a single mixture of divided substance with
undivided form; and this mixture of two forces,
the same and the different, became the origin of
motion; which indeed was not accomplished in the
easiest way, being extremely difficult.

Now all these proportions are combined harmon-
ically according to numbers; which proportions

were scientifically divided according to a scale
which reveals the elements and the means of the
soul's combination. Now seeing that the earlier
is more powerfále in power and time than the l-
later, the deityndid not rank the soul after th
substance of the body, but made it older, by
taking the first of unities, 384 (12x16). Knowi:
this first, we can easily reckon the double and
the triple; and all the terms together, with ti
complements and eighths, must amount to 114,69⁻
and likewise the divisions (sum of the tone-se
uences of 36 tones, amounting to 384x27, the po...
fect cube).

4. Planetary Revolutions and Time.

God the eternal, the chief ruler of the Uni-
erse, and its creator is beheld alone by the mi:
but we may behold by sight all that is producec
this world and its parts, howmany soever they a
in heaven; which, as being ethereal, must be di
ided# into kinds, some realting to sameness, c
others to difference. Sameness draws ĥhtward a
that is withánt, along the general eastward mov
ment from the west. Difference draws from with:
all self-moved portiens from west to east, fo:
tously rolling around and along by the superio:
power of sameness.

The different's movement, being divided in
harmonical proportion, assumes the order of se:
circles. Nearest to the earth, the Maon revolv
in a month; while beyond her the Sun completes
his revolution in a year. Two planets run a co:
equal with that of the Sun: Mercury, and Juno,
also called Venus and Bucifer, because shepher
and people generally are not skilful in sacre
tronomy, confusing the western and eastern ri:
The same star may shine in the west when follc
the Sun at a distance great enough to be visib
in spite of solar splendor; and at another ti:
in the east, when, as heral of the day it ris
before the Sun, leading it. Because of its run
together with the sun, Venus is Lucifer freque
but not always; for there are planets and st:

OF ANY MAGNITUDe seen above the horizon before
sunrise, herald the day. But the three other
planets, Mars, Jupiter and Saturn have their
peculiar velocities and different years, completing their course while making their periods
of effulgence, of visibility, of obscuration and
eclipse, causing accurate rising and settings.
Moreover they complete their appearances conspic-
uously in east or west according to their position
relative to the Sun, who during the day speeds
westward, which during the night it reverses, under the influence of sameness; while its annual
revolution is due to its inherent motion. In resultance of these two kinds of motion it rolls
out a spiral, creeping according to one portion,
in the time of a day, but, whirled around under
the sphere of the fixed stars, according to each
revolution of darkness and day.

Now these revolutions are by men called portions of time, which the deity arranged together
with the world. For before the world the stars
did not exist; and hence there was neither year,
nor periods of seasons, by which this generated
time is measured, and which is the representation
of the ungenerated time called eternity. For as
this heaven has been produced according to an
eternal pattern, (the world of ideas, -- so according to the pattern of eternity was our world-time
created simultaneously with the world.
    5

5. The Earth's Creation by Geometric Figures.

The Earth, fixed at the centre, becomes the
hearth of the gods, and the boundary of darkness
and day, producing both settingsnand risings,
according to the occultations produced by the
things that form the boundary, just as we improve our sight by making a tube with our closed
hand, to exclude refraction. The Earth is the
oldest body in the heavens. Water was not produced
without earth, nor air without moisture; nor could
fire continue without moisture and the materials
that are inflammable; so that the Earth is fixed

upon its balance as the root and base of all other
substances. Of produced things, the substratum is
Matter, while the reason of each shape is abstract
Form; of these two the resultance is Earth, and
Water, Air and Fire.

This is how they were created. Every body is com
posed of surfaces, whose elements are triangles;
of which one is right-angled, and the other has all
unequal sides, with tue greater angle thrice the
size of the lesser; while its least angle is the

the third of a right angle, and the middle on is
the double of the least; for it is two parts out
of three; while the greatest is a right angle,
being one and a half greater than the middle one,
and the triple of the least. Now this unequal side
triangle is the half of an equilateral triangle,
cut into two equal parts by a line let down from
the apex to that base. Now in each of these triang
there is a right angle; but in the one the two sic
about the right angle are equal, and in the othe.
all the sides are unequal. Now let this be called
a scalene triangle; while the other, the half of
the square, is the principle of the constitutien
of the Earth. For the square produced from this
scalene triangle is composed of four half-squares
and from such a square is produced the cube, a bo
the most stationary and steady in every way; havi
six sides and eight angles, and on this account i
Earth is a body the heaviest and most difficult t
be moved, and its substance is inconvertible, bec
befause it has no affinity with a triangle of ano,
kind. Only the Earth has as peculiar element the
square and this is the element of the three othe
substances, Fire, Air and Water. For when the ha
triangle is put together six times, it produces a
solid equilateral triangle; the exemplar of the
amid, which has four faces with equal angles, wh
is the form of Fire, as the easiest to be moved
and composed of the finest particles. After this
ranks the octohedron, with eight faces and six a
angles, the element of Air, and the third is the
acosahedron, with twenty faces and twelve angles
the element of Water, composed of the most numero
and heaviest particles.

These the n, as being composed of the same
element, are transmuted. But the deity has made
the dodecahedron, as being the nearest to the spher
sphere, the image of the Universe. Fire then, by
the fineness of its particles, passes through all
things; and Air through the rest of things, with
the exception of Fire ; and Water through the
Earth. All things are therefore full, and have no
vacuum. They cohere by the revolving movement of
the Universe, and are pressed against, and rubbed
by, each other in turn, and produce the never-fail
failing change from production to destruction.

6. Concretion of the Elements.

By making use of these the deity put together
this world, sensible to touch through the particles of Earth, and to sight through those of
Fire; which two are the extremes. Through the
particles of Air and Water he has conjoined the world
world by the strongest chain, namely, proportion;
which restrains not only itself but all its subjects. Now if the conjoined object is a plane surface, one middle term is sufficient; but if a
solid, there will be need of two. With two middle
terms, therefore, he combined two extremes; so
that as Fire is to Air, Air might be to Water,
and Water to Earth; and by alternation, as Fire
is to Water, Air might be to Earth; and by inver-
sion, as Earth is to Water, Water might be to Air,
and Air to Fire; and by alternation, as Earth is to
to Air, so Water might be to Fire. Now since all
are equal in power, their ratios are in a state of
of equality. This world is then one, through the
bond of the deity, made according to proportion.
Now each of these substances possesses many
forms; Fire, those of Flame, and Burning and
Luminousness, through the inequality of the triang
angles in each of them. In the same manner, Air
is partly clear and dry, and partly turbid and fogg
foggy; and Water partly flowing and partly con-
gealed, according as it is Snow, Hoar-frost, Hail
or Ice; and that which is Moist, is in one respect

flowing as honey and oil; but in another is compact, as pitch and wax; and of compact-forms there
are some fusible, as gold, silver, copper, tin,
lead and steel; and some friable, as sulphur, pith
pitch, nitre, salt, alum, and similar metals.

7. Composition of the Soul.

After putting together the world, the deity plan
planned the creation of living beings, subject to
death, so that, himself being perfect, he might
perfectly work it out according to his image.
Therefore he mixed up the soul of man out of the
same proportions and powers, and after taking the
particles and distributing them, he delivered them
over to Nature, whose office is to effect change.
She then took up the task of working out mortal
and ephemeral living beings, whose souls were draw
drawn in from different sources, some from the
Moon, others from the Sun, and others from various
planets, that cycle within the Difference, — with
the exception of one single power which was deriv
derived from Sameness, which she mixed up in the
rational portion of the soul, as the image of wisdom in those of a happy fate.
Now of the soul of man one portion is rational
and intellectual; and another irrational and unin-
tellectual. Of the ligical part the best portion
is derived from Sameness, while the worse comes
from Difference; and each is situated around the
head, so that the other portions of the soul and
body may minister to it, as the uppermost of the
whole tabernacle. Of the irrational portion, that
which represents passion hangs around the heart,
while desire inhabits the liver. The principle of
the body, and root of the marrow, is the brain,
wherein inheres leadership; and from this, like
an effusion, through the back-bone flows what remains, from which are separated the particles for
seed and reason; while the marrow's surrounding
defences are the bones, of which the flesh is the
covering and congealment. To the nerves he united
joints by ligatures, suitable for their movement.

Of the internal organs, some exist for the
sake of nourishment, and others for safety; of
communications, some convey outside movements to
the interior intelligent plahes of perception,
while others, not falling under the power of appre-
prehension, are unperceived, either because the
affected bodies are too earth-like, or because
the movementsnare too feeble; the painful movements tend to arouse Nature, while the ppeasurable
lull Nature into remaining within itself.

8.   Sensations.

Amongst the senses, ###### the deity has in
us lit sightto view the objects in the heavens,
and for the reception of knowledge; while to
make us capabłe to receive sp eech and melody,
he has in us implanted hearing, of which he who
is deprived thereof from birth will become
dumb, nor be able to utter any speech; and that
is why this sense is said to be related closest
to speech.

As many affections of the body as have a name
are so called with reference to touch; and others
from relation to their seat. Touch judges of the
properties cornected with life, such as warmth,
coldness, dryness, moisture, smoothness, roughness, and of things, that they are yoelding,
opposing, hard, or soft, Touch also decides of
heaviness or lightness. Reason defines these
affections as being centripetal and centrifugal:
which men mean to express when they say below,
and middle. For the centre of a sphere is below,
and that part lying above it and stretching to
the circumference, is called upwardness.

Now what is warm appears to consist of fine
particles, causing bodies to separate; while
coldness consists of the grossness of the particles, causing a tendency to condense.

The circumstances connected with the sense of
taste are similar to those of touch. For substances grow either smooth or rough by concretion
and secretion, by entering the pores, and assum-

.ing shapes. For those that cause the tongue to melt away, or that scrape it, appear to be rough; while those that act moderately in scraping appear brackish; while those that inflame or separate the skin are acrid; while their opposites, the smooth and sweet, are reduced to a juicy state.

. Of smelling, the kinds have not been defined; for, because of their percolating through narrow pores, that are toos stiff to be closed or separated, things seem to be sweet-smelling or bad-smelling from the putrfaction or concoction of the earth and similar substances.

A vocal sound is a percussion in the air, arriving at the soul through the ears; the pores (or communications) of which reach to the liver; and among them is breath, by the movement of which hearing exists. Now of the voice and hearing, that portion which is quick is acute; while that which is slow, is grave; the medium being the most harmonious. What is much and diffused, is great; what is little and compressed, is small; what is arranged according to musical proportions is in tune; while that which is unarranged, and unproportionate, is out of tune, and not properly adjusted.

The fourth kind of things relating to the senses is the most multiform and various, and they are called objects of sight, in which are all kinds of colors, and an infinity of colored substances. The principle are four: white, black, brilliant (blue) and red, out of a mixture of which all other colors are prepared. What is white causes the vision to expand, and what is black causes it to contract; just as warmth expands, and cold contracts, and what is rough contracts the tasting, and what is sharp dilates it.

9.   RESPIRATION

It is natural for the covering of animals that live in the air to be nourished and kept together by the food being distributed by the veins through the whole mass, in the manner of a stream, conveyed as it were by channels, and moistened by the breath, which diffuses it, and carries it to the ·

EXTREMITIES. Respiration is produced through there
being no vacuum in nature; while the air, as it flows
flows in, is inhaled in place of that which is ex-
haled, through unseen pores, such as those through
which perspiration-drops appear on the skin; but
a portion is excreted by the natural warmth of
the body. Thennit becomes necessary for an equivalent portion to be reintroduced, to avoid a vacuum, which is impossible, for the animal would bo
longer be concentrating, and single, when the covering had been separated by the vacuum.

Now in lifeless substances, according to the
analogy of respiration, the same organization occurs. The gourd, and the amber, for instance,
bear resemblance to respiration.

Now the breath flows through the body to an
orifice outwards, and is in turn introduced
through respiration by the mouth and nostrils,
and again after the manner of the Euripus, is
in turn carried to the body which is extended
according to the expiration. Also the gourd,
when the air within is expelled by fire, attracts
moisture to itself; and amber, when the air is separated from it, received an equal substance. Now
all nourishment comes as from a root from the heart;
and from the stomach, as a fountain; and is conveyed
to the body, to which, if it be moistened by
more than what flows out, there is said to be
an increase; but if less, by a decay; but the point
of perfection is the boundary between these two,
and is considered to exist in an equality of efflux
and influx; but when the joints of the system are
broken, should there no longer exist any passage
for the breath, or the nourishment not be distributed,
then the animal dies.

10. DISORDERS.

There are many things hurtful to life, which
are causes of death. One kind is disease. Its beginning is disharmony of the functions, when the simple
powers, such as heat, cold, moisture or dryness
are excessive or deficient. The come turns and
alterations in the blood, from corruption, and the
deterioration of the flesh, when wasting away,

should the turns take place according to the changes, to what is acid, or brackish, or bitter, in
the blood, or wasting away of the flesh. Hence
arise the production of bile, and of phlegm, diseased juices, and the rottenness of liquids, weak
indeed, unless deeply seated; but difficult to
cure, when their commencement is generated from
the bones, and painful, if in a state of inflammation of the marrow. The last of disorders are
those of the breath, bile and phlegm, when they
increase and flow into situations foreign to them,
or into places inappropriate for them, by laying
hold of the situation, belonging to what is better,
and be driving away what is congenial, they fix
themselves there, injuring the bodies, and res-
olving them into the very things.

These then are the sufferings of the body;
and hence arise many diseases of the soul; some
from one faculty, and some from another. Of the
perceptive soul the disease is a difficulty of
perception, of the recollecting, a forgetfulness
of the appetitive part, a deficiency of desire
and eagerness; of the affective, a violent suffering and excited madness; of the rational,
an indisposition to learn and think.

But of wickedness the beginnings are pleasures and pains; desires and fears, inflamed by
the body, mingled with the mind, and called by
different names. For there loves and regrets,
desires let loose, and passions on the stretch,
heavy resentments, and appetites of various kinds,
and pleasures immoderate. Plainly, to be unreason-
ably disposed towards the affections is the limit
of virtue, and to be under their rule is that
of vice; for to abound in them, or to be superior to them, places us in a good or bad position.
Against such impulses the temperaments of our
bodies is greatly able to cooperate, whether
quick or hot, or various, by leading us to mel-
ancholy or violent lewdness; and certain parts,
when affected by a catarrh, produce itchings
and forms of body more similar to a state of
inflammation than one of health; through which
a sinking of the spirits and a forgetfulness,
a silliness and a state of fear are witnessed.

11. Discipline.

Important, too, are the habits in which persons are trained, in the city or at home, and the:-
daily food, by luxury enervating the soul, or fo
tifying it for strength. For the living out of
doors, and simple fare, and gymnastic exercises,
and the morals of companions, produce the greatest effect in the way of vice and virtue. These
causes are derived from our parents and the elements, rather than ourselves, provided that on our
part there be no remissness, by keeping aloof from
acts of duty. The animal cannot be in good condition unless the body possesses the better properties under its control; namely, health and
correct pasception, and strength and beauty. Now
the principles of beauty are a symmetry as regards its parts, and as regards the soul. For
nature has arranged the body, like an instrument,
to be subservient to, and in harmony with, the
subjects of life. The soul must likewise be brought
into harmony with its analogous good qualities,
namely, in the case of temperance, as the body is
in the case of health; and in that of prudence,
as in the case of correct perception; and in
that of fortitude, as in the case of vigor and
strength; and in that of justice, as in the case
of beauty.

Nature, of course, furnishes their beginnings;
but their continuation and maturation rsult from
carefulness; those relating to the body, through
the gymnastic and medical arts; those to the soul,
through instruction and philosophy. For these
are the powers that nourish and give a tone to
the body and soul by means of labor and gymnastic
exercise, and pureness of diet; some through
drug medication applied to the body, and others
through discipline applied to the soul by means
of punishments and reproaches; for by the encouragement they give strength and excite to an on-
ward movement, and exhort to beneficial deeds.
The art of the gymnasium trainer, and its nearest
approach, that of the medical man, do, on application to the body, reduce their powers to the

utmost symmetry, purifying the blood, and equal-
izing the breath, so that, if there were there
any diseased virulence, the powers of blood and
breath may be vigorous; but music, and its leader, philosophy, which the laws and the gods ordained as regulators for the soul, accustom, per
persuade and partly compel the irrational to oby
obey reason, and the two irrational, passion and
desire, to become, the one mild, and the other
quiet, so as not to be moved without reason, nor
to be unmoved when the mind incites either to
desire or enjoy something; for this is the definition of temperance, namely, docility and
######### firmness. Intelligence and philosophy
the highest in honor, after cleansing the soul
from false opinions, have introduced knowledge,
recalling the mind from excessive ignorance,
and setting it free for the contemplation of
divine things; in which to occupy oneself with
self-sufficiency, as regards the affairs of a
man, and with an abundance, for the commensurate period of life, is a happy state.

12. Human Destiny.

Now he to whom the deity has happened to assign somewhat of a good fate, is, through opinion, led to the happiest life. But if he be mor-
ose and indocile, let the punishment that comes
from law and reason follow him; bringing with it
the fears ever on the stretch, both those that
originate in heaven or Hades; how that punishment
ments inexorable, are below laid up for the unhappy, as well as those ancient Homeric threats
of reataliation for the wickedness of those defiled by crime(Odyssey, xii:571-599). For as we
sometimes restore bodies to heath by means of
diseased substances, if they will not yield to
the more healthy, so if the soul will not be led
by true reasoning, we restrain it by false.
Strange indeed would those punishments be called
since, by a change, the souls of cowards enter
into bodies of women, who are inclined to in-
sulting conduct; and those of the blood-stained

WOULD BE PUNISHED BY BEING INTROduced into the
bodies of wild beasts; of the lascivious, into
the bodies of sows and boars; of the lightminded
and frivolous into shaper of aeronautic birds;
and of those who neither do learn or think of
nothing, into the bodies of idle fish.

On all these matters, however, there has, at
a second period, been delivered a judgment by
Nemesis, or Fate, together with the avenging d ...
deities that preside over murderers, and those
under the earth in Hades, and the inspectors of
human affairs, to whom God, the leader of all,
has intrusted the administration of the world
which is filled with gods and men, and the rest
of the living beings which by the demiurgic
creator according to the best mode of an unbegotten, eternal and mentally-perceived form.